The Appraisal of Mineral Rights

with emphasis on minerals as real property

By

Terrel Shields

2015 ©

Table of Contents

Table of Contents. ... 2

Chapter I - A History of Oil Valuation. ... 4

Chapter II - Definition of the Mineral Right. ... 13
    The "Bundle of Sticks". ... 13
    The Mineral Deed. ... 19

Chapter III - The Lease. ... 25
    The Interest. ... 26
    The Header. ... 28
    The Consideration. ... 28
    The Term. ... 29
    Royalties. ... 30
    Expiration. ... 31
    Amendments. ... 31
    The Bottom Line. ... 37
    Cold Drafting. ... 37
    What about not leasing?. ... 38
    After the Lease. ... 40
    Pooling. ... 40
    Division Orders. ... 42

Chapter IV - Apportionment & Unit Rule. ... 45
    The Unit Rule. ... 45
    Apportioning Value. ... 47
    Examples of the Unit Rule. ... 50
    Risk Adjustments. ... 51

Chapter V - Discounting Partial Interests. ... 53
    Quick Review of Fractional Interests. ... 53

Chapter VI - Methods & Strategies. ... 57
    Approaches. ... 57
    Decline Curves and Reserve Estimates. ... 57
    Including Mineral Rights in the Value. ... 66
    Speculative Value. ... 68

Chapter VII - An Engineering Valuation. 71
    Engineering Value v. Market Value. 71

Chapter VIII - Gathering the Data. 74

Chapter IX - Methods for Valuing Mineral Rights. 75
    Minerals without Production. 75
    Producing Property - No Reserves Estimate Methods. 78
    Producing Property - Reserves Based Methods. 80

Chapter X - Anatomy of the Play. 87
    The Boom - Bust Cycle 87
    Dealing with the Rapidly Changing Market. 88

Appendix A - Engineering-Style Report 90
    Future Production. . . . . demonstration report only - 91
    Gas Prices. . . . . demonstration report only - 92
    Royalty Interest. . . . . demonstration report only - 92
    Conclusions. . . . . demonstration report only - 94

Appendix B - Fracking & Earthquakes. . . . . Fracking - 98

About the Author

    Terrel Shields is an Arkansas based consultant. He is a registered geologist in Arkansas and certified general appraiser in both Arkansas and Oklahoma. He is also a mineral owner. After spending 10 years in various positions from hydrocarbon logger to Operations Manager for Exploration Service, Inc., Midland ,TX, he started a consulting firm specializing in wellsite geology and hydrocarbon logging.

    After doing some consulting to banks who held mineral rights in the late 1980s, the idea of becoming an appraiser was born. It was compelling both in starting in the appraisal business as the oil business waned and certification became a requirement for appraisers. It was only natural to eventually return to the oil business valuing mineral rights and this book is the outcome of years of finding few resources on minerals to serve the appraiser, real estate professional, or owner who encounters oil and gas issues in the course of their own work.

Chapter I - A History of Oil Valuation
"And the rock poured out for me streams of oil! - Job 29:6b"

From ancient times when Job lamented the ease of his past life, until the modern era, oil was recognized as an important product useful for everything from sealing boat hulls to burning for fuel. Indeed, most plastic products today are oil-based and the refining techniques developed for oil have been modified to develop new products from renewable products like corn and soybean oils. The Seneca Indian skimmed oil from the creeks of New York and Northwest Pennsylvania, as did other natives from coast to coast. They often used it as medicine. The modern era of petroleum exploitation was initially financed by a drug company. Col. Edwin Drake had no clue that oil would be the engine that would transform the Industrial Age into the Oil Age.

From the early days of Col. Drake's discovery in Titusville, PA, people recognized the value of oil[1] and bought and sold oil rights in complex deals. Often it was the landowner who got the short end of the stick, as they say. Some things don't change. Professor William Wright, only six years after Drake's discovery wrote a book about the oil fields of Pennsylvania[2] in which he describes a method of valuing oil that is still used today.

The method Wright described was the sort of income-multiplier that appraisers know so well. In the most refined method, that price per barrel was based on the price of oil times $1,000. If a well made 100 barrels of oil in one day and oil wells were selling for $6,000 per barrel, then the well sold for $600,000.

Promoters often tried to sell their wells at the peak of production, early after production commenced. Even then these early oil barons realized that the production was unlikely to remain

---

[1] Please note that "oil" "petroleum" and "gas" are used rather loosely within this book and refers to liquids, solids and gases in some instances. In specific instances where differentiation is necessary, the distinction will be made explicitly.

[2] William Wright : "The Oil Regions of Pennsylvania", Harper Brothers, (1865), reprinted by Elibron Classics (2005.)

steady but would decline rapidly with time. If the decline was too rapid, the investor stood a very good chance of not getting their money back. And the price could change. Wright reported the typical price then was $3,000 to a winter peak of $5,000 dollars per barrel/day with the better crude oils bringing even more. It was also recognized that the quality of the oil makes a difference in price. In those earliest years, it was kerosene that was preferred. Gasoline was a "waste" product often simply dumped into a creek and set on fire.

Drake was hired by a pharmaceutical company seeking to sell oil as a salve and commonly sold as Seneca Oil named for the local native Indian tribes. That name was later corrupted to "snake oil" and such homeopathic products can still be found in drug stores today. Within months of Drake's discovery, and despite oil being found a wonderful substitute for whale lamp oils, the price of rock oil fell from $1 per gallon ($42 per barrel) to less than half that.

Drake himself developed a number of innovations including casing, but in the end, he went bankrupt and died a pauper. The citizens of Titusville paid for his memorial next to the graves of he and his wife. One problem arose from the first wells. He had promised the landowners a fixed fifty cents per gallon. When prices fell, Drake was quickly underwater, with the royalty being well below the market price of the oil. But as Wright reports, royalty was already commonly based upon 1/8th of the well production by the end of the Civil War.

The oil industry can be said to have saved the whales from extinction. Whales had become scarce and whale oil, popular for lighting of oil lamps, was expensive. Petroleum undercut the price and idled no small number of whale ships. Many of the whalers sought jobs in the booming oil business. They were familiar with rigging sails, climbing rigging, and the very term, "oil rig" denotes the influences of the displaced seaman. Derrick, knowledge box, dog house, etc. are terms that were imported from the sailors now employed as oil drillers. Roughneck, roustabout, and tool pusher are terms that seem to have originated in southeast Texas. Many terms common to rigs were developed in Texas around the time of the Spindletop[3] discovery.

As noted above, valuers of oil set the price of oil property and the current form of leasing land and giving the landowner (mineral owner) a portion of the proceeds developed early. Also, developed early was the concept of the law of capture[4]. The law of capture meant that since oil is not defined nor confined by property boundaries, the first straw in the soda often got the lion's share of the oil. By the end of the Civil War the idea of the landowner getting one-eighth of the proceeds was already common practice.

Production Always Declines

Even by 1865 early engineers and geologists recognized that wells decline in production over time. These decline curves were plotted month by month and were used to predict when a property would cease producing economic quantities of petroleum. These were rarely straight line declines, rather were hyperbolic declines, declining more rapidly in the first years, then declining at a slower rate as time passed. A decline curve is simply a plot of production (monthly total MCF or BO [5]) vs time. The area under the curve using simple calculus (assuming any calculus is simple) reflects the total production to date whereas the projected curve area is the remaining reserves. Alternatively, you can plot cumulative

---

[3] Spindletop in Texas was the initial discovery where high pressure blowouts were first a problem and the location of the first rotary well drilling methods.

[4] The law of capture applied to oil, water, and wild animals. The ownership of these mobile resources were owned by the one who captures them on their property even if it originated on an adjacent parcel. This was old English common law practice. To avoid being drained by an adjacent landowner, many leases require a leaseholder to defend and protect the property by drilling an adjacent well to prevent drainage.

[5] MCF is the volume of gas at room temperature is 1000 cubic feet. Barrels of Oil (BO) is a volume measurement. A barrel is 42 gallons. See comment in footnote page 87. This is not the same as the 55 gallon chemical drum common today.

production and in that case the bow of the curve is upwards. And in most cases, as oil production slowed, water production increased and hastened the end of the well's profitable life.

Early petroleum engineers plotted curves and made some empirical guesses about how much oil or gas remained in the ground. Paul Paine [6] described oil valuation techniques that were simple plots of predicted oil. He also described methods similar to Wright's days of payout multiple as well as other common metrics. He also notes that the time value of money affects the value of oil still in the ground.

But all the writing about valuation generally reflected valuations that are for oil companies and the landowner's position is noted in passing only. Even today, there seems to be very little information and valuation expertise available to the royalty owner. The petroleum engineer is much in demand today for valuing reservoir properties and typically charge a much higher fee than the average landowner is willing to pay. Perhaps the harried appraiser should take note of that when pricing their services.

Mineral Valuing - The Real Estate Step-child

There have been two major sources of mineral-related information that is valuable to the appraiser today. The first is John Gustavson. Gustavson is retired as of this writing but the company, Gustavson and Associates is under the leadership of Edwin Mortiz, a geologist and appraiser in Colorado. Gustavson was the first person to attempt to meld the ideas of classical appraisal with that of the engineering profession in the valuation of mines, minerals, and oil and gas. Gustavson recognized the large discounts necessary to make reserve estimates match actual transaction values. The risk involved in buying and selling mineral rights reflects more than a simple time value of money adjustment. It involves considerable risk.

Professor John Baen, a professor of Real Estate at North

---

[6] Paul Paine : "Oil Property Valuation", John Wiley & Sons (1942)

Texas University, has written several articles on mineral rights valuation and is recommended reading[7] for all appraisers dealing with the valuation of mineral rights. He offers a number of strategies for valuing mineral rights and addresses the issues of the environment in some of his writings. He also recognizes that oil and gas is valued in ways not typical of other classes of real estate and that using the same methods that the industry does seems to result in values that are more sensible and realistic. Understanding the characteristics of the mineral interest is necessary to value them properly.

Some terms:

Probabilistic- Deterministic:

> Deterministic methods refer to solving a problem with a set of known data or inferred data such as a prior history of production, whereas probabilistic refers to statistical methods of estimating what data is or should be. It is a projection or speculation about the problem. Engineering estimates are probabilistic usually while market value is usually a deterministic endeavor.

Normative economics:

> Normative economics refers to judgment based decision. What is the price something should be rather than what is actually is. Its weakest is that humans are unpredictable and often react on rumor or fear rather than facts.

Positive economics:

> Positive economics is based on facts and uses a minimum of hindsight or future projections to estimate worth. Its weakness is failure to identify or quantify an intrinsic value to an item.

---

[7] Dr. John S. Baen : "Oil and Gas Mineral Rights in Land Appraisal" Appraisal Journal, April 1988 ; "The Impact of mineral rights and oil and gas activity on Agricultural Land Values", Appraisal Journal, Jan. 1996 ; are two of a number of articles written by Dr. Baen

Equivalence

When a well produces both oil and gas, sometimes instead of reporting both, an operator will report in either gas equivalence or oil equivalence. The BTU content of a thousand cubic feet of gas (which is written as MCF) is about one-eighth that of a barrel of oil. That varies with the quality of the gas and the oil but generally eight to twelve is used as a multiplier. 8,000,000 cu. ft. of gas is 8,000 MCF, and therefore is equal to 1,000 barrels of oil.

Likewise, 100 barrels of oil equals 800,000 cu. ft. or is equal to 800 MCF. These terms can be easily mixed up with production figures which estimate the flow in Barrels of oil per day and MCF per day.

BOE – Barrels of Oil Equivalent
MCFGE – Thousand Cu. Ft. Gas Equivalent

Approximately 8 times the energy is in a barrel of oil as in a 1,000 CF (MCF) of gas. When gas is $8 then oil should be $64/bbl. When that ratio begins to get "out of whack", companies begin to switch to the cheaper fuel. So a problem exists then when the price differential is stretched from the actual BTU content. For years, the two were generally in sync, but with the unconventional plays and the demand for oil high while natural gas is in a serious glut, the ratio based on price may be 20 or more. Many companies use 12 as a hedge, being less than the price ratio, and more than the BTU ratio.

To look good on paper, a company with a high ratio of gas to oil, may still report BOE instead of MCFGE because it makes it appear that the value of the reserves are higher than they actually are. The best strategy is to apply a methodology that is consistent and segregates the oil from the gas.

The valuation of minerals rights is an on-going problem for appraisers and this treatise is purposed to provide the appraiser with the basic methods available to them. With the dearth of tried and true methods from within the appraisal industry, reaching out to the industry for help is absolutely necessary. For too long the appraiser has been a footnote to the engineering methods applied by oil companies. This book in meant to fill in some of the gaps between the industry methods and the methods available to the appraiser.

Finally, appraising is an art form which depends upon the heuristic (judgment) of common sense, rules of thumbs and traditional methods. Those methods include the cost approach, the sales approach, and the income approach. The engineering methods applied by the industry are generally probabilistic where projections are made and using those projections an outcome is predicted. That may involve proprietary information such as seismic and well logs that are unavailable to the appraiser. It may involve advanced math such as the Monte Carlo Simulation to run scenario analysis and rank prospects. Again, these are tools that are not available to the typical appraiser of mineral interests in fee simple.

The appraiser, on the other hand, is more accustomed to deterministic methods. Determinism relies upon historic data such as production records to value the mineral right and attempts to translate income into value on a present day basis. Over-application of forecasts renders the analysis academic and introduces a high amount of uncertainty into a process that is already high risk (uncertain).

And as noted above, the appraiser may not have access to the most critical data available to the operator of a well. The royalty owner is likewise as the mercy of the operator and leaseholder and we should never lose sight of the fact that the mineral owner lacks executive control over his property. The valuation of such passive interests also must analyze the impact upon the land surface whether the mineral owner is the landowner or whether his interest is severed from that of the surface owner.

There is no class of people who enter a contract with others who are less prepared to defend their rights than the mineral owner, particularly the landowner who retains or owns mineral rights. And there is no class of resources more difficult to value and to understand than are mineral rights.

While this treatise is mostly about oil and gas since those interests are the most broad-based, common, and expansive mineral estates, we will not be discussing much about quarry rocks, mining, or construction materials. But many of the same factors apply to those mineral interests as well as oil and gas.

The commonality of minerals and petroleum is that to value the overall property, you must value the reserves. Reserves are simply the amount of oil and gas, or of the mined material (quarry rock, ore, etc.) From the reserves you can develop a value in place and allocate that value to the various interests in the project. The value will depend upon the state of development and the remaining reserves. A proposed mine reserves can be valued by drilling test holes, then estimating the volume of rock or ore that can be mined. The quality of that ore will determine the net value in situ.

The same applies to oil an gas. Petroleum resides in the pore spaces of rock. As depth increases, this petroleum is compressed and deep gas wells hold a much higher volume of gas than would a shallow formation. The deep gas would be liquid under pressure and require less space. But ultimately, the volume is controlled by the amount of pore space and open fractures between the grains of clay, lime, or sand in the reservoir rock. This volume can be estimated. And with an estimate of volume we can make judgments about the value of the mineral therein.

Part I - Mineral Rights

Chapter II - Definition of the Mineral Right
"First define the problem"

The Bundle of Rights

You can

Sell

Lease

Use

Give away

Enter (egress and ingress), or

Refuse to do any of the above

This is the old SLUGER acronym appraisers are familiar with from basic appraisal classes. The fee simple includes the bundle of rights for both surface and minerals as well as any other unencumbered right.

Each "stick" can be sold, leased, used, given way, egress/ingress, or refuse to do anything INDEPENDENT of the other sticks in the Fee Simple Bundle. Some say that mortgaging or encumbering with an easement is also a right of a "stick" holder.

It is possible to own all of the rights in a parcel of real estate or only a portion of them. A person owning all of the rights is said to have fee simple title. Fee simple title is regarded as an estate without limitations or restrictions. Less-than-complete estates result from partial interests that are created by selling, leasing or otherwise limiting the bundle of rights in the fee estate. An appraisal assignment may require the appraisal of fee simple title or any partial interest such as the leasehold interest of an easement or a mineral right.

The "Bundle of Sticks"

In the bundle of rights theory, ownership of real property is compared to a bundle of sticks. Each stick represents a distinct and separate right, which can be individually sold, leased, entered, given away, or to choose to exercise more than one or none of these rights. Although subject to certain limitations and restrictions, private enjoyment of these rights is guaranteed by law under the U. S. Constitution.

Real Estate

Real estate is the physical land and appurtenances including structures attached thereto. Real estate is immobile and tangible. Legally defined, real estate includes land and all things that are a natural part of it (e.g. trees and minerals) and all things that are attached to it by people (e.g. building and pavement.)

Fee Simple

Fee Simple title to property includes all the rights, air, surface and minerals. *Cuius est solum, eius est usque ad coelum et ad inferos*, meaning from heaven to hell. Anything less than that is something less than Fee Simple

Disclaiming the mineral right from consideration means you are appraising the "Fee in Surface". You don't have that check box on the URAR. You will need to explain the actual rights appraised in the addendum. For the appraiser, your issue is to correctly identify the rights being appraised. In many cases, those rights may have long been severed. In other words, the owner may not even have the mineral rights and therefore, does not have "fee simple." And if the owner has and yet has leased to an oil company that fee simple is encumbered with a lease. Banks are getting smarter about this mineral interest and demanding they be named in the lease, get the proceeds from the lease, etc.

In the depression, many people with mortgages on their property were still able to sell off their mineral rights without the bank objecting. As a result, people surrendered the farm in foreclosure but sold the mineral rights, often to get enough money to move to California. Where such rights were still intact, there were bankers who severed the interest and bought those themselves. Grocery men who took the mineral deeds for outstanding bills.

One in western Oklahoma told that his own father in law was concerned that he would go bankrupt and threatened to whip him over the issue. He argued that if he didn't take the mineral deed then these people were going to go on anyway and he would get nothing. Years later he retired with a number of royalty checks that made him a wealthy man. The Kennedy family, among many others, accumulated mineral deeds during the depression for $1 per acre or less and created a small oil company to handle the royalty holdings. Much of it was accumulated as payment for goods sold by the Kennedy's.

The fee simple can be sold as a unit or can be divided and a "severed" interest sold. The mineral rights may be retained or deeded away by the owner of the fee simple.

A Key Point Concerning Minerals as Real Property

Production (oil, gas, coal, etc.) is personal property once it leaves the ground. The Mineral Right has a value based upon the Principle that Value is the anticipated future benefit. It may not have ANY oil or gas. It may be too expensive to drill. It is Anticipation of a future benefit that sets value.

Further, Mineral rights are dominant rights. The surface is subservient to mineral rights and this is an enormous bone of contention with landowners. Conflicts are inevitable. The appraiser should be clear upon what they are including or not including in the value. This is a scope of work issue.

Note that there are two valuation perspectives. One is that of the landowner. In it, we are generally dealing with a real property interest under Standards I and II of Uniform Standards of Professional Appraisal Practice (USPAP). The interest therefore does not include well head equipment, storage tanks, etc. Those are personal property allocated to the operator of the well.

The working interest is not composed of a real property interest entirely. While it does include the leasehold interest, it also is the sum of the equipment (personal property) value and as working interests includes an intangible business enterprise value. A third factor is that of carried interests which are not part of the leasehold nor the leased fee, and are strictly intangible interests. These are often called "Overriding Royalty Interests (ORRI)" and are a form of payment to those who perform services to the operator in lieu of a direct payment.

Is Determining Mineral Value Different From What You Do?

Farm land sells not for the tangible benefit of owning a chunk of ground. It sells based upon the "anticipation" that it will grow crops. Recreational property sells for the anticipation of pursuing ducks, deer, etc. Houses sell for the anticipation that it is more profitable to own rather than lease. Mineral rights sell for the anticipation of finding economic amounts of oil and gas, coal, etc. not for any known quantity that can be proved or measured accurately.

How to Read a Deed

Mineral rights are often held by reservations in a deed. They commonly read like this:

> The NW¼ of the NE¼ of Sec. 13 - Twp. 23 N - Rge 8 E, Indian Meridian, less and except all rights to the oil, gas, or coal herein.

Or,

> The NW¼ of the NE¼ of Sec. 13 - Twp. 23 N - Rge 8 E, Indian Meridian, subject to any reservation of record.

When encountering such language in a deed it is not a bad policy to request an explanation from the client. Legalese and liability has reduced the legal description to using less specific language when attempting to identify things that are not part of the rights being conveyed.

With so many old deeds and difficult to trace histories, people are often surprised to find they own less than they thought they were buying. A lot of people find their mineral rights were withheld by reservation many years ago leaving them with "fee in surface" only.

Warranty Deeds or Mineral Deeds actually convey the deeded interest to another party. Like this:

Notice that the deed explicitly refers to the property as "real" property. MINERAL RIGHTS ARE REAL PROPERTY RIGHTS!

> :......convey unto Grantees' heirs and assigns forever, the following real property lying in Woodruff County, Arkansas:
>
> AN UNDIVIDED ONE-HALF (½) INTEREST IN ALL THE MINERALS, INCLUDING, BUT NOT LIMITED TO, OIL, GAS, AND OTHER

The appraisal forms do not have check boxes for fee in surface, rather give the choice of leasehold or fee simple. Fee simple, in reality, is a rare bird these days. Many deeds have restrictions, certain uses that the property cannot be put to, or are encumbered by access easements or utility easements which limit the true unfettered use of a parcel.

Without the mineral interest, the proper oil field terminology from landmen is usually "fee is surface" but "fee simple less mineral right" or similar language is equally descriptive. We will leave it to the academics to make a more definitive delineation. You will almost certainly encounter language used where the fee simple is described as fee simple but does not include mineral rights.

The issue of describing the fee simple also must wrestle with the situation where the mineral rights are leased or held by production. In this case while the surface be an unencumbered fee interest, the mineral right is now a leased fee and therefore is once again something less than fee simple.

On the following pages are examples of a mineral deed and a reservation of a mineral interest in a warranty deed for land (surface). When such language exists it may be difficult to determine if the mineral is actually severed or the legal is meant to cover all bases when the history isn't clear. But when mentioned explicitly in a mineral reservation, it is quite likely there is a reservation of record. Further research would need to be done.

Title companies relied upon insurance now. They rarely check for deed reservations and easements beyond a thirty to fifty year time frame. Many mineral interests were severed during the 1920s and especially in the 1930s with the depression. People raised money by selling the mineral right just to keep from losing the farm, or to pay off people that they owed money to.

Note that there are documentary stamps on these deeds. Oklahoma and Arkansas are among the few states where a mineral deed requires a documentary stamp (also known as deed or revenue stamps). More on the subject of deed stamps later.

The Mineral Deed

Deed Stamps (a. k. a. - Documentary Stamps, Revenue Stamps, etc.) in this jurisdiction are $3.30 per thousand, thus the sales price is $92.40 ÷ 3.3 = 28 x $1,000 = $28,000, or $2,029 per net mineral acre (13 net mineral acres).

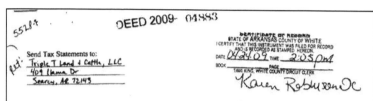

The Mineral Reservation

Many minerals are reflected in mineral reservation. Typically, all the mineral rights are reserved. In some deeds only the oil and gas rights are reserved. Others may name the product actually reserved, or included in the sale.

Below a deed simply states that the surface conveys but is unclear about where the minerals are. Were they conveyed previously, or are they being retained with this deed?

> THAT Bill G. Rush, a single man, party of the first part, in consideration of the sum of Ten Dollars ($10.00) and other good and valuable consideration, the receipt of which is hereby acknowledged does by these presents grant, bargain, sell and convey unto Martin Livestock, LLC, an Oklahoma Limited Liability Company, party of the second part, its successors and assigns, all of the following described real property and premises situated in Ottawa County, State of Oklahoma, to-wit:
>
> The NE ¼ SE ¼ and N ½ NW ¼ SE ¼ of Section 31, and W ½ NW ¼ and NW ¼ SW ¼ and S ½ NE ¼ and E ½ NW ¼ and NW ¼ NE ¼ of Section 32, All in Township 29 North, Range 23 East of the Indian Meridian, Ottawa County, Oklahoma, according to the Government Survey thereof LESS AND EXCEPT a tract beginning at the Northeast corner of the NW ¼ NE ¼ of said Section 32; thence West 575 feet; thence South 1895 feet; thence East 1895 feet; thence North 575 feet to the Northeast corner of the SE ¼ NE ¼; thence West 1320 feet; thence North 1320 feet to the point of beginning.
> AND
> ➔ SURFACE RIGHTS ONLY IN AND TO: The East 575 feet of the NW ¼ NE ¼ of Section 32, Township 29 North, Range 23 East of the Indian Meridian, Ottawa County, Oklahoma LESS AND EXCEPT THE North 210 feet of the East 555 feet of said NW ¼ NE ¼.

A straightforward reservation

> KNOW ALL MEN BY THESE PRESENTS:
>
> THAT, PARKER DEAN ROBERTS AND E. FAYE ROBERTS, husband and wife, parties of the first part, in consideration of the sum of ONE DOLLAR and other valuable considerations, in hand paid, the receipt of which is hereby acknowledged, do hereby grant, bargain, sell and convey unto WESLEY R. KOEHN AND ALICE I. KOEHN, husband and wife, R. R. # 1 Box 87, Inman, Ks 67546, as joint tenants and not as tenants in common, with the right of survivorship, the whole estate to vest in the survivor, parties of the second part, the following described real property and premises situate in DELAWARE County, State of Arkansas, to wit:
>
> All that portion of the SE¼ SE¼ of Section 18, Township 21 North, Range 25 East, lying North and West of existing section line roads containing 40 acres, more or less, Delaware County, Oklahoma.
>
> Less and except the oil, gas, and other mineral rights having heretofore being reserved of record.

Define the Problem

One serious problem is determining the exact mineral acreage associated with the subject. You always want the simple situation of 100% minerals intact, but frequently you find that the ownership is less than 100%, often 50% and on rare occasion an owner has retained mineral rights to an adjacent property where he sold off the surface interests.

Without title opinion or mineral opinion, the appraiser should caveat the report with an extraordinary assumption that you assume the mineral estate is intact and you are therefore appraising "fee simple" without reservation.

Mineral Rights are the Dominant Estate

Egress and ingress of the surface is usually required for a site to drill, explore, or exploit the mineral right. Conflicts arise over the issue. And many conflicts originate in the observation that most people with took little thought in consideration of the mineral right when purchasing a rural property. Unless the buyer was careful to discover the omission, they rarely thought about except in areas where on-going exploration has piqued the interest of the entire community. Recently many multi-listing services have added a line for minerals. Buyers in North Dakota, Ohio, Pennsylvania, and other parts of the United States had never considered the implication of actual development in their area. Some were thought to be "played out" years ago. Old areas of production are now being revived. Old mineral rights are being sold after being dormant for decades. And some development has occurred in areas that were never developed in the first place. The Fayetteville Shale is typical of a play that blossomed in areas the old geologists thought to be a "driller's graveyard."

When landowners discover they do not have the right to keep drillers off their property, conflict is inevitable. In the early 1980's near Sallisaw, Oklahoma, 2 security guards were killed by an angry landowner. The guards were providing security for an oil company which had a confrontation with the owner and the sheriff had escorted the rig onto the premises. The landowner was convicted of second degree murder.

In North Texas in 1989, the author was working on a drilling rig when a renter pulled a gun on our toolpusher as he entered the property to check a water pump on a pond leased by the drilling company. Ironically, the renter was months behind on the rent. The landowner evicted her and her horses while she was incarcerated in the local jail over the incident. Many people in the Rocky Mountains have purchased land that the government never sold with the mineral right. Homestead laws were changed to retain the right to all but oil and gas, then later changed to reserve all the mineral rights to the Federal Government.

Who Owns What?

The Bureau of Land Management (BLM) routinely holds auctions of mineral interests controlled by the federal government. Land can be nominated for the sale by the general public. After environmental studies are complete, the BLM may include such land in a sale. The minimum bid is $2 an acre.

If the mineral owner decides to exploit the mineral estate will the lack of control affect the surface owner's property value? It can. If a well is drilled, crops can be damaged. Fields bisected by a road may have to be planted differently. Irrigation systems may be affected. Noise may be a problem to the homeowner.

The value of the surface can be impacted, with or without mineral rights. An appraiser has to be able to determine these impacts and make adjustments to the fee simple to arrive at the true market value of the surface property. The absence of mineral rights does not mean the end of the appraiser's responsibility to measure the impact this has on the value of the property being appraised.

So Who Are Your Peers?

USPAP requires we do what our peers would do with a similar assignment. Who are your peers? Other clueless appraisers? Petroleum Engineers? Something else? A market based valuation is not an engineering assignment.

In developing an appraisal where mineral rights are involved, the appraiser must educate themselves on the subject or they must obtain expert assistance. When dealing with other resource appraisals the question of competence and expertise is the first one that needs to be answered.

The expert hired must understand the appraisal process and the goal that you have. Valuations that are based on reserves forces the appraiser to rely upon an estimator and USPAP allows the appraiser to rely upon such information with the only caveat being that the appraiser "have a reasonable basis for believing that those individuals performing the work are competent." That means you have relied upon an estimator who has experience and who is trained to analyze well production figures. That may be a geologist, an engineer, a member of the American Institute of Mineral Appraisers (AIMA), the Society of Petroleum Evaluation Engineers (SPEE), or other recognized organization in a related field.

Summary

Surface rights can be appraised without the mineral rights included. But the impact of mineral rights or lack thereof must be "considered" and "analyzed" (USPAP requirement) because it is a condition that can affect the value of the rights being appraised.

The appraiser should report that fact after considering the situation and assessing the likelihood that the property value will be impacted by the presence or absence of an intact mineral estate.

Environmental issues relating to development include road issues, traffic, noise, and the visual effect of development activity. And counter-impacts include a general wealth effect as jobs are created, landowners are paid for water, gravel or building materials, or citizens take jobs on drilling rigs and wells.

In areas where earthquake activity has increased, the appraiser needs appropriate language to disclaim any liability regarding same. This is particularly important in areas where actual damage has been reported, such as south-central Kansas, northeastern Ohio, central and north-central Oklahoma, Faulkner County, Arkansas, etc.

It would not be a bad policy for homeowners in these regions to carefully photography with dates all exterior and interior walls to document any damage that might occur after that date.

If you encounter the mineral interest and must value it, be sure to obtain the kind of advice or assistance that the Competency section of USPAP requires.

Chapter III - The Lease
"Look before you Lease - Jim Stafford"

When a landowner agrees to lease their mineral rights, the fee simple becomes complex, not simple. Suddenly, the fee simple is divided into a surface component and a subsurface component. The surface remains in the hands of the owner (assuming it isn't leased, too.) The subsurface becomes a leased fee estate if the mineral rights are leased.

Further, mineral rights can mean coal, diamonds, gold, precious stones, as well as oil or gas. Often these will be explicitly excluded from the Oil & Gas Lease (OGL). Even then, if limited to hydrocarbons, then carbon dioxide, helium, nitrogen, and other gases may be present. These may be a nuisance and impact the quality of the gas, or may be very valuable. Hydrogen sulfide gas, in particular, is a toxic and dangerous one. It often impacts the quality of the petroleum and must be flared or separated from the produced gases.

It is not a bad practice to explicitly state what minerals are being leased and if the impact of those other minerals (gold, coal, etc.) is potentially an issue. The author was once contacted by a title insuring company to estimate the value of the minerals under a property that was in a shallow gas producing area where only a few wells existed. But once I went into investigating the situation, I decided to withdraw from the assignment.

The situation was that a buyer had hired the local title company to determine that the mineral rights were intact on a 400 acre parcel. They concluded they were. They were mistaken. There was 80 acres in the midst of the parcel that had went into foreclosure in the past and the governmental lender had retained the mineral rights upon selling off the surface property. The buyer uncovered the error and sued. The title insurer then attempted to make it good by buying the minerals under the 80. Being a governmental entity, of course, they denied the request. The suit thus proceeded.

The problem was that the buyer was a quarry company who had test-drilled the 400 acres and determined that a large limestone body was under it. They intended to mine the property for stone, not oil and gas. Clearly, the title company was hoping for a low appraisal reflecting the oil and gas production potential only. The value was in the stone. Without the test holes, it would have been speculative (probabilistic) whether the limestone was valuable or not. With the core test holes, it was no longer speculative rather a deterministic estimate of the economic potential could be made.

The Interest

The lease holder gets the larger share because of the high cost of finding, drilling, and producing oil or gas. The lessor can negotiate with the lessee for the best royalty and the highest bonus money to be paid. Every part of a lease agreement is negotiable but one item is constant. Namely, if oil or gas is struck, the lease is held in perpetuity until such time as oil and gas ceases to be produced. It is said to be "held by production" or HBP.

The Lease

An oil and gas lease differs little from commercial leases except the terms and terminology are different. Market rents are the typical returns of a similar well and contract rents are the actual returns. Leases are contracts and as such differ little from other commercial leases in the actual intent of the lease.

Terminology

Lease

A written document in which the rights of use are transferred by the owner to another party for a specified period of time

Leased Fee

Represents the owners position, the lessor who receives a Royalty if production is found

Leasehold

Represents the property under tenure of lease. The lessee or tenants position

Reversion

The right of the property owner to receive the property back after the expiration of the lease

Term

The length of time the lease is good for

Bonus Payment

Up-front money tendered at the signing of a lease

Delay Rental

If drilling does not commence in the first year, a delay rental payment is made. Usually a few dollars per year

Royalty

Royalty is the percentage of the proceeds that the mineral owner will be paid out of any production. Gas is paid in money from the gross sale of the gas, whereas oil is given to the mineral owner as a share of the oil, but it is usually marketed by the oil company unless the mineral owner desires to sell his oil independent of the Petroleum company

The Header

Producers 88
Paid-up, Pooling
OIL, GAS AND MINERAL LEASE

This Lease Agreement (the "Lease") is entered into this _____ day of _____, 2006, (the "Effective Date") between _____ as "Lessor," whether one or more, and Tiger Ted, LLC a Texas Limited Liability Company, 00 Davenport Road, Justin, Texas 76247 , as "Lessee."

The Producers 88[8] label is meaningless and was the original printing shop number for the form. "Paid Up" means the annual delay rentals are paid up at the time a bonus is paid. "Pooling" provisions mean that the lessor agrees to be pooled voluntarily. Pooling, integration, unitization all describe a process of putting more than one tract together as a single operating unit. All the parties within a pool earn the royalty pro-rated with all the other parties. This means whether a well is actually on your property or not, you will get your share of the proceeds so long as it is in the unit.

This can get very complicated with "Cross-sectional" wells. These wells with very long laterals may allocate a portion of a well between two integrated units.

The Consideration

In consideration of the sum of Ten Dollars ($10.00) and other valuable consideration, the receipt of which is acknowledged, and of the royalties and agreements of Lessee contained in this Lease, Lessor grants, leases, and lets exclusively to Lessee, its successors and assigns, all of the land described in this Lease, together with any reversionary rights of Lessor, for the purpose of

---

[8] The term Producers 88 was created in the 1920s by a Tulsa printing company. They kept stock supplies of oil and gas leases and the old Producers Oil Co. was a client. It was the 88th form in the book and so when Producers or anyone else wanting a similar form asked for them to be printed up, they asked for the "Producer's 88" form and the company printed off the number desired.

exploring by geological, geophysical, and all other methods, and of drilling, producing and operating wells or mines for the recovery of oil, gas and other hydrocarbons, and all other minerals or substances, whether similar or dissimilar, that may be produced from any well or mine of the leased premises, including primary, secondary, tertiary, cycling, pressure maintenance methods of recovery, and all other methods, whether now known or unknown, with all related incidental rights, and to establish and utilize facilities for surface and subsurface disposal of salt water, and to construct, maintain and remove roadways, tanks, pipelines, electric power and telephone lines, power stations, machinery and structures thereon, to produce, store, transport, treat and remove all substances described above, and their products, together with the right of ingress and egress to and from the land subject to this Lease and across any other land now or later owned by Lessor. The land that is covered by and subject to this Lease is situated in JACKSON COUNTY, ARKANSAS, and is described as follows and referred to in this Lease as the "land" or the "lands".

<lease description>

This section defines the amount paid and the legal description of the tract leased.

The Term

1. Without reference to the commencement, prosecution, or cessation at any time of drilling or other development operations, or the discovery, development, or cessation at any time of production of oil, gas, or other minerals, and notwithstanding anything else contained in this Lease to the contrary, this Lease shall be for a term of **Three (3)** years from the date stated above (the "Primary Term") and as long thereafter as oil, gas, or other minerals are produced from the lands, or land with which the lands are pooled, or as long as this Lease is continued in effect as otherwise provided by the terms of this Lease.

Option to Renew

Lessee is hereby given the option, to be exercised prior to the date on which this Lease or any portion thereof would expire in accordance with its terms and provisions, of extending the Lease for a period of **three (3)** years as to all or any portion of the acreage then held hereunder which would expire unless so extended, the only action required by Lessee to exercise this option being the payments to Lessors of the additional consideration of the sum of **$300.00** per net mineral acre for each acre so extended. If this Lease is extended as to only a portion of this acreage then covered hereby, Lessee shall designate such portion by a recordable instrument.

Renewals are problematic. If they don't drill in 3 years, why expect them to drill in the next three years. If leases are worth less, they won't renew. If they are worth more, then they can renew the lease for less than market. This is a provision that is common, but a savvy royalty owner will strike the clause.

Royalties

2. The royalties to be paid by Lessee are: (a) on oil, and on other liquid hydrocarbons saved at the well, seventeen percent **(17%)** of that produced and saved from the land, the same to be delivered at the wells or to the credit of Lessor in the pipeline to which the wells may be connected, Lessor's interest in either case shall bear its proportionate share of any expenses for treating oil to make it marketable as crude; (b) on gas, including casing head gas or other gaseous substances produced from the land and sold on or off the premises, seventeen percent **(17%)** of the gross proceeds at the well received from the sale of gas, provided that on gas used off the premises or by Lessee in the manufacture of gasoline or other products, the royalty shall be the market value at the well of seventeen percent **(17%)** of the gas so used, and as to all gas sold by Lessee under a written contract, the price received by Lessee for that gas shall be conclusively presumed to be the gross proceeds at the well or the market value at the well for the gas so sold; (c) on all other minerals mined and marketed, ...

This sets the percentage of production earned by the lessor, which in turn is central to valuing the mineral rights being held for both leaseholder and leased fee interest.

Expiration

3. If at the expiration of the Primary Term of this Lease, oil, gas, or other minerals are not being produced from the lands or land pooled with the lands subject to this Lease, but Lessee is then engaged in drilling or reworking operations, this Lease shall remain in force so long as drilling or reworking operations, are prosecuted (whether on the same or different wells) with no cessation of more than sixty (60) consecutive days, and if they result in production, so long thereafter as oil, gas, or other minerals are produced from the lands or land pooled with the lands subject to this Lease. If production of oil, gas, or other minerals on the lands or land pooled with the lands should cease from any cause after the primary term, this Lease nevertheless shall continue in force and effect as long as additional drilling operations or reworking operations are conducted on this Lease, or on any acreage pooled with the lands, which additional operations shall be deemed to be had when not more than sixty (60) days elapse between the abandonment of operations on one well and the commencement of operations on another well, and if production is obtained this Lease shall continue as long as oil, gas, or other minerals are produced from the lands or land pooled with the lands, and as long thereafter as additional operations, either drilling or reworking are had on the lands or pooled lands.

"Held by Production" means the lease is not terminated so long as gas or oil is produced. That could be decades or even a century or more. The anniversary date of the lease is not a "drop-dead" date. Most courts allow a lease to be extended so long as operations have commenced, often as little as building a cattle guard into the site, or construction thereof. The mineral owner is well served to have a "drop-dead" clause inserted into the lease so that substantial activity has to be on-going and activity pursued with a drill rig large enough to complete the well.

Amendments

4. (a) There is reserved and excepted from this lease and reserved to Lessor(s), their heirs, successors or assigns, all lignite, coal, uranium and metallic ores and it is understood and expressly provided that the terms "mineral" or "minerals", "other mineral" and "other minerals" shall refer to oil, gas and other hydrocarbons

and their respective constituents associated with the production of oil and/or gas and shall not refer to or include lignite, coal, uranium and metallic ores or other substances not associated with the production of oil and/or gas.

Landowner Protections

(b) Lessee agrees to keep all its equipment on the land in a neat and orderly condition, and generally to maintain the land in a neat and orderly and attractive condition. Lessee shall consult with Lessor with respect to the type, size and location of all routes of ingress and egress and roads, and the location of all pipelines, drilling sites, production sites and any other operations conducted on the land prior to such operations. Reasonably in advance of conduction drilling operations upon the land, Lessee shall furnish Lessor written notice stating the nature of such operations and designating that portion of such land to be used in connection therewith.

Setbacks and Damages

(c) It is understood that the land is now being, or may hereafter be, used by Lessor as owner of the surface estate thereof, or the tenants or lessees of Lessor (Lessor or such tenants being herein sometimes referred to as "surface owner"), for farming, ranching, and hunting operation, and Lessee shall, in its operations hereunder, interfere as little as reasonably possible therewith. Lessee shall not locate any well drilled hereunder within four hundred (400') of the residences or barns, if any, on the land, nor within two hundred feet (200') of any other building now or hereafter situated upon the surface of the land, without the prior written consent of surface owner, Lessee shall pay all damages directly or indirectly caused by its operations hereunder to timber, grass, growing crops, livestock, game birds or animals, water wells, fences, roads, building, or other strictures or improvements, and any and all other property, whether real or personal, of surface owner. Lessee shall bury all pipelines below plow depth, but in no event less than three feet (3') below the surface.

Some states are applying or considering setback notifications to anyone living within a set distance of a drilling rig.

Pooling Agreement

UNITIZATION
Lessee is hereby granted the right to pool and unitize the Marcellus Shale, Onondaga, Oriskany or deeper formations under all or any part of the land described above with any other lease or leases, land or lands, mineral estates, or any of them whether owned by lessee or others, so as to create one or more drilling or production units. Such drilling or production units shall not exceed 640 acres in extent and shall conform to the rules and regulations of any lawful governmental authority having jurisdiction in the premises, and with good drilling or production practice in the area in which the land is located.

In the event of the utilization of the whole or any part of the land covered by this lease, lessee or designated operator shall before or after the completion of a well, record a copy of its unit operation designation in the County wherein the leased premises is located, and mail a copy thereof to the lessor. In order to give effect to the known limits of the oil and gas pool, as such limits may be determined from available geological or scientific information or drilling operations, lessee may at any time increase or decrease that portion of the acreage covered by this lease which is included in any drilling or production unit, or exclude it altogether, provided that written notice thereof shall be given to lessor promptly.

As to each drilling or production unit designated by the lessee, the lessor agrees to accept and shall receive out of the production or the proceeds from the production from such unit, such proportion or the royalties specified herein, as the number of acres out of the lands covered by this lease, which may be included from time to time in any such unit, bears to the total number of acres included in such unit.

The commencement drilling, completion of or production from a well on any portion of the unit created under the terms of this paragraph shall have the same effect upon the terms of this lease as if a well were commenced, drilled, completed or producing on the land described herein. In the event, however, that a portion[9] only of the land described in this lease is included from time to time in such a unit then proportionate part of the delay rental, hereinafter provided, shall be paid on the remaining

A good Clause to Have

(d)   Lessee agrees to make a one-time damage payment to Lessor for each drill site on the land, payable in advance of commencing preparation. THIS LEASE DOES NOT AFFECT THE SPECIFIC TERMS OF ANY SURFACE DAMAGE SETTLEMENT. IT IS UNDERSTOOD THAT ANY SURFACE DAMAGE SETTLEMENT(S) PERTAINING TO THESE LANDS WILL BE NEGOTIATED SEPARATELY.

Livestock protections

(e)   Lessee will not cut or otherwise obtain access through or over any fence or fences situated on, or on any boundary of, the land in connection with any operations of the land without first obtaining the written consent of surface owner or its representative. Lessee agrees promptly after making such cut to install and maintain a cattle-guard and gate (of a type and design reasonably satisfactory to surface owner), and said cattle-guards and gates shall become the property of surface owner. If necessary, during drilling operations, all pits, well and wellhead equipment, tanks compressors, separators and all other surface installations and equipment of Lessee shall be fenced to prevent the exposure of live stock to any hazardous condition, and such fences shall be maintained by lessee in good condition and repair.

Limiting Uses by Employees

---

[9]This is a royalty owner unfriendly clause as part of your property could lay outside the unit and not bring in income while at the same time, being held by production and the lease terms and preventing the owner from leasing to another party. It literally creates a buffer zone around a unit and ultimate could drain the undeveloped portion of the lease rendering it worth less if not worthless.

(f) Neither Lessee, its successors or assigns, not its employees, contractors, agents or representatives, or anyone on the land with the permission of or at the invitation of Lessee, its successors or assigns, shall bring firearms onto the land or be permitted to hunt or fish on the land.

Top Leasing Provisions

Top leasing refers to the practice of leasing a property before the original lease expires. There is an element of risk for the company doing the leasing in that a company may enter the property at the last minute and commence drilling which renders a top lease void. However, this is usually only done with the company is confident that the expiring lease is not likely to see a drill rig in time to save the lease.

Some people try to insert a clause which allows them to match any offer made by a competitor during the last year of the lease even when it is clear the original lessee will not be drilling a well. But the question is why lease back to someone who isn't going to drill a well?

Owner Friendly Clauses

The lessor generally is presented with a oil company friendly lease. There are lease clauses that favor the lessor. These are often difficult to negotiate but when you have the upper hand or prices are high in the region, you are more likely to get these inserted in the lease.

Vertical & Horizontal Pugh Clause - If a well has multiple zones that can produce but the operator elects only to produce one zone, then the rest of the reserves are not making the mineral owner any money and won't until the operator decides to drill another well to the undeveloped oil or gas, or re-complete the well in that zone. While they might not produce from the zone, the operator can still "book reserves". That is they can claim the value of the undeveloped zone and even borrow money against it. That does the property owner little good.

To protect themselves, savvy mineral owners apply a "Pugh Clause" named after a lawyer who first tried this. Each Pugh clause is different and there is no set language. A vertical "Pugh" relates to those zones above and below the producing zone and simply says that after the end of the primary term, these zones cannot be held by production from other zones. The language often is something like "at the end of the lease shall expire for all zones 100' above and below the Mississippi Lime...".

The Horizontal Pugh clause can be even more important. Many landowners with large holdings often in several counties were surprised to find that a single well drilled on any part of their property held all the other acres, regardless how far away, by production. This mean ranches of 1,000 or more acres may have only 1 well and very little income for decades.

Favored Nations Clause -Such a provision (strangely named as it is) provides that if any lessor owning mineral interests within a given area, unit, or within a certain distance within a certain time is paid a higher bonus or royalty, the operator will agree to match that offer and make up the difference. This is a useful clause where the owner is not sure of the value. But perhaps its greatest use is as due diligence for a trustee of an estate. That trustee could be an accountant, financial advisor, or an heir or child of the trust holder.

When a trustee is given control of the leasing their due diligence is to get the best royalty and bonus available. A disgruntled sibling or heir could claim that the trustee was not competent and seek their removal, or claim damages. The Favored Nations Clause gives the trustee some protection from that claim.

No Post-Production Expenses - No clause has caused more difficulty for mineral owners. Prior to 1992, few leases had such clauses. Until deregulation of the natural gas industry, a flat percentage of the market value at the wellhead was sent to the royalty owner. The only deductions were for taxes which are often required to be paid up front from either the operator or the first buyer (pipeline or refinery).

After deregulation companies were charged transportation

costs from the well head to the pipeline. The operator frequently owned that pipeline and compressor and thus charged themselves and the royalty owner excessive fees. Additionally, clauses often gave the operator free gas to run the compressors, all at the expense of the royalty owner. These deductions may run over 40% and rise with the reduction in production to the point that they barely break even. If they fail to at least break even, then there is a case to be made that the economic life is over and the lease is no longer held by production.

One particular onerous post-production deduction is "marketing". No one seems to really know what that means or can explain that well. What is marketing? What is to market? Gas sells itself pretty much.

Even when post-production is prohibited, another clause called "enhancements" can be a catch all that results in the same outcome. The company keeps a part of the mineral owners money. And courts are divided on the issue but generally agree that costs necessary to market the gas can be deducted even when a no post-production clause is in force.

The Bottom Line

All leases are contracts. No lease is "fixed" and immovable. Companies may be but a contract is a contract no different from any other. A landowner may be negotiating from a position of weakness or strength, but they can still refuse to sign. That has implications in states with pooling. You can be force pooled. But you have options even then.

Leases are complex. People who take several days to decide on what pickup to buy and read every line in the sales contract when buying a house may not understand anything about the contract they signed that just gave control of their mineral rights to strangers. This is the time to seek professional help or to educate yourself on mineral rights. After you sign, it is too late.

Cold Drafting

No, this is not about beer....

Promoters, flim-flam artists, and conmen often jump into a play. They may offer very high bonuses on a 90 day or 120 day bank draft. They get the owners to sign the leases and they take them and the landowner holds a draft which is supposedly going to be made good within the 90 - 120 day period. Before that period ends, the landowner expects the draft to be paid. Frequently, it is not. The promoter is attempting to sell your signed leases to oil companies during the 90 day period. Only then will the draft be made good. The lease is fungible and even after the draft expires, if the promoter manages to sell your leases to someone, they can simply send you the money or make the draft good.

If the landowner's draft is not made good, bad things can happen. First, the promoter could simply file the leases anyway. It would be upon the landowner to go to court and contest the lease. Even if the lease is never filed, the landowner needs to go through a legal process to get their title cloud removed or needs to obtain the actual lease signed and destroy it.

Also, it can create a lot of ill-will at your bank. If the draft comes back unpaid, the mineral owner will still owe the bank for the processing - which can be several dollars. And they are unlikely to shoulder that expense themselves. Angry customers often vent at the bank, but drafts are only as good as the person who writes you a draft. And anyone who writes you a 90 day draft isn't likely to be a legitimate representative of any oil company. They are what is known as a promoter.

What about not leasing?

Some royalty owners hope to 'cash in' by converting their interest into a Working Interest or by participating in a well if integrated (pooled, unitized, etc.). Such neophyte "oil tycoons" quickly find that an unscrupulous operator can "operate them to death" by charging hokey expenses against their interest. Such small minority interests usually preclude the party from seeking legal help because the cost exceeds the value of the small interest. "Transparency" in the cost accounting is problematic and investors need to be wary of most operators, especially small obscure companies or companies with "history".

Even in legitimate deals landowners who operate or convert their interest into a working interest are taking a big risk and will be a subordinate party to the operator. As such, they likely will not have any executive control over the operation and will have to be extraordinarily lucky to have a happy outcome. With each new well, productive or not, you will be asked for your portion of the drilling costs up front. If you fail to provide it, then a 400% penalty or worse, may be applied - depending upon state law. You can count on getting little, if any of your mineral value with such a penalty.

Investing in wells is a business for deep pockets. And deep pockets do not invest all their money in a single well or drilling unit. They tend to spread their investment out over a larger area with various operators so as to reduce risk. If you think you really want to do that, then extract out a single acre from the legal and lease the rest. If you have a few thousand dollars to play with, you will be able to see how these play out.

With a single acre in a 640 acre drilling unit, you will owe about 1.5% of the total cost. For an AFE (authority for expenditure - the cost estimate) of $4,000,000 you will owe about $6,000 before the well is drilled. After the well is drilled additional completion money will need to be tendered. Then within six months after the first production you will get your pro-rated share of the proceeds. If it is a dry hole, you get to deduct the expense as intangible drilling cost (IDC) from taxes. With congress considered this a loophole and many politicians hoping to close it, consider the tax implications of not being able to deduct the cost.

While on taxes, most investments allow for deductions. Investing in buildings, machinery, etc. assume that this expense can be depreciated over the life of the item. The depletion allowance is no different. Mineral rights are wasting assets. After all the oil and gas is produced, the owner is left with nothing. It has depreciated and royalty owners can deduct 15% from their gross mineral income as a standard rate. There are people in congress who want to eliminate that as well. Regardless your opinion of the IDC or oil companies, removing the depletion allowance would be grossly unfair to royalty owners.

After the Lease

Once leased, the mineral owner waits for the company to make the first move. If they decide to drill on your property, you will get notice and surface damages can be negotiated. Again, provisions in the lease will influence the location of the roads and the wells. Under no circumstances should a landowner allow the operator to use your property to access other wells or property. So again, watch the contract they want signed. Read it.

Since the lease likely has exploration language, you are allowing them to enter your property to run a seismic survey or other exploration prior to the selection of a drilling site. The site may be on land within the pooled unit you belong and yet not be on your land.

Pooling

Pooling (also known as integration, unitization, etc.) is the process by which mineral owners are pooled together into a single operating unit. Pooling may be voluntarily usually by the lease provisions. And 39 states have some form of forced pooling or integration law. West Virginia and Pennsylvania have pooling provisions that do not apply to the Marcellus Shale. These states, while early oil producers, had seen little drilling since the 1920s and their laws and regulations are ill-suited for the modern world and many need seriously updated.

In states with the Public Land Surveying System (PLSS) was applied, drilling units are generally equally sized and based on section - township - range descriptions. 640 acres is a common drilling unit (or spacing) In old oil wells, 40 or 80 acre units may apply. An 80 that is ½ mile to the north-south, and ¼ mile east-west is called a "stand up" 80 and one that lies ½ mile east-west by ¼ mile north-south is a "laydown" 80. Since oil and gas know no real boundaries except geological, these units are contrived and artificial but do attempt to fairly distribute the production among the parties involved.

The larger the unit, the fairer it is since it distributes the production over larger areas. Some very large, often special situation units, are 4,000 or even 100,000 acres in size. One that comes to mind is the McElmo Dome unit in Southwest Colorado where carbon dioxide is found and sent to West Texas to improve production of old oil wells in the Permian Basin. By larger scale pooling and measured field development, the landowner will get smaller checks but those checks will tend to change little from month to month until a new well is added, and in the process average out the income stream over a longer period of time, perhaps several generations. In the case of the McElmo Dome above, production initiated with about 15 wells and two to four are added annually. The field has been active for over thirty years.

In metes and bounds states (mostly east of the Mississippi) the units are irregular usually but may be on the same order of scale as those of the PLSS states, often being 600 to 1200 acres in size in the unconventional plays.

Once pooled, each well in a drilling unit will need to have a division of interest or division order created to identify what part of the well is your interest. This is done for every one, both working interests and royalty interests.

Division Orders

A division order is sent to the mineral owner to confirm their interest in a well. Each well drilled in a unit will provide just such a division order. The mineral owner should keep them in a safe place. They are not recorded in a courthouse. These are often flawed and the interest should be checked closely. Here is an example why.

This clearly shows that the acreage is 40 acres in a uniform 640 acre section. The lease terms were one-eighth royalty.

GIVEN
640 acre spacing or drilling unit and the owner signs a lease with a 1/8 royalty

40 acres ÷ 640 acre unit * 1/8th = 0.007813

0.00585938 is exactly 75% of 0.007813, in other words, the

owner is being paid for 30 acres.

Example of Well Plots below

When such errors are caught years afterwards, there is no

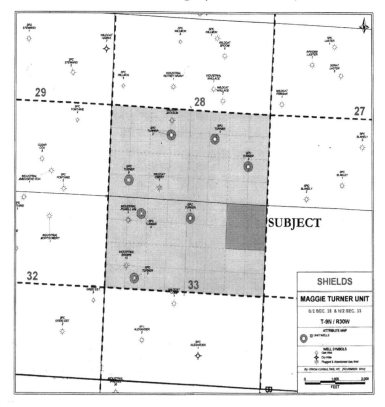

recourse except to sue and then expect only to get the last few years reimbursed because the statutes of limitations will cut off anything beyond that. It was detectable with the most casual of math, and should have been corrected almost 40 years earlier.

    Things get very complex when crossing unit boundaries. In that case, the production from a well has to be pro-rated from the percentage of the well lying in its respective unit. And some wells impact up to four sections (units.) In those cases, the DOI (division of interest) is different for each well, and may result in multiple checks or DOI's to deal with in the check stub. To add to the confusion, some states (Arkansas comes to mind) may allow the operator to pass the production figures to the various working interests with leases and they, in turn, pay the royalty, or pay all the "excess" royalty, which is the royalty above the minimum $12\frac{1}{2}\%$ (1/8th) mandated by most states. In Arkansas it is not unusual to have three checks for each well each month. A large landholder had 101 wells impacting his property (about 1200 acres total in four counties) and needed a full time bookkeeper to handle the influx of checks each month.

Chapter IV - Apportionment & Unit Rule

"...never to regret and never to look back. Regret is only for wallowing in."–
K. Mansfield

Std. 1 - 4 (USPAP)

"(e) An appraiser must analyze the effect on value, if any, of the assemblage of the various estates or component parts of a property and refrain from valuing the whole solely by adding together the individual values of the various estates or component parts.

Comment: Although the value of the whole may be equal to the sum of the separate estates or parts, it also may be greater than or less than the sum of such estates or parts. Therefore, the value of the whole must be tested by reference to appropriate data and supported by an appropriate analysis of such data."

A similar procedure must be followed when the value of the whole has been established and the appraiser seeks to value a part. The value of any such part must be tested by reference to appropriate data and supported by an appropriate analysis of such data"

The above is reinforced in the Uniform Standards for Federal Land Acquisition (USFLA, a. k. a. - The Yellow Book). USFLA requires appraisers working for federal projects to apply the "Unit Rule".

also from USPAP,

"(f) An appraiser must analyze the effect on value, if any, of anticipated public or private improvements, located on or off the site, to the extent that market actions reflect such anticipated improvements as of the effective appraisal date."

The Unit Rule

The market value concept adopted by the courts to be applied in federal acquisitions generally requires application of the so-called unit rule. This rule has two aspects; one relating to the interests, or estates, into which ownership of real estate may be carved, and the second relating to the various physical components of real estate.

The first aspect of the unit rule requires that property be valued as a whole rather than by the sum of the values of the various interests into which it may have been carved, such as lessor and lessee, life tenant and remainderman, and mortgagor and mortgagee, etc. This is an application of the principle that it is the property, not the various interests, that is being acquired.

"The whole property or unit valuation remains applicable even where the ownership is divided between such inherently diverse interests as surface rights and timber rights or surface and mineral rights. That does not necessarily mean, however, that the independent values of the various interests are not admissible in a condemnation trial; but if they are admitted it is for the sole purpose of aiding the trier of fact in fixing the value of the property as a whole." Likewise, it is not inappropriate for appraisers to consider the independent values of the interests, but again, only for the purpose of better estimating the market value of the whole property.

Value the Whole Property

To value the fee simple estate the elements of comparison should include those items affecting the rights appraised. Mineral rights are impacting the local real estate market and differences in the mineral right of the subject and the comparables can be used as a basis for making an adjustment.

The temptation to combine a value for the surface and a value for the mineral separately, then add them together creates the potential of summation of the estates being inaccurate. To do it this may be necessary but the results almost always over-values the fee simple. 1 + 2 3, in this case.

As counter-intuitive as it is, remember the buyer of the surface is not the same pool of buyers as the buyers of mineral rights and each doesn't necessarily want to buy or even know how to value the estate of the other. Intact mineral rights (still part of the original fee simple) rarely sell to professional investors, but relates to the local agricultural market. Thus the market participants are different from those who participate in the market for mineral rights. Which Market are you appraising for?

How Do Buyers Differ?

The buyer of mineral rights will look at the income stream and/or geology and make a determination of the value accordingly. Comparable sales in the area are weighed less and in no case would the sophisticated buyer of minerals rely solely upon sales prices rather than their own opinion of its potential future income (benefit.)

The buyer of land with minerals may allow value for the mineral interest as a passive income source. Comparables sales are all they usually have to work with. Few such buyers would pay for a geological investigation or engineers report. Rumor is the most common information on which they base offers. They are also motivated by desire to avoid losing control of the surface. If you own the surface, integrated by forced pooling or not, you can keep the drilling rig off your property or limit it to certain parts of the property.

Apportioning Value

Paired Sales

Comparable sales unaffected by the mineral interest must be adjusted to account for the additional value of the mineral right. The appraiser must test these adjustments to see that they are reasonable. This is nothing more or less than the appraiser would do to make an adjustment to a dwelling based upon an estimate of differences in square footages or differences in acreage.

Methods of Adjustments[10]

The following are Quantitative methods:
Paired data set analysis (Paired Sales)

---

[10] From the Appraisal of Real Estate, 12th Ed., Appraisal Institute, Chicago, Illinois

Statistical analysis
Graphic analysis
Trend analysis
Cost - related analysis
Secondary data analysis
The following are Qualitative methods
Relative comparison
Ranking analysis
Personal interviews (polling)

Because the cost related method is applicable only to improvements, and there is usually insufficient data to make a statistical or graphic analysis, the most applicable methods are paired sales extraction, trend analysis, and personal interviews (polling,) in addition to income methods.

In paired sales analysis, partitioning of the mineral value of a fee simple or the sale of a severed mineral interest gives an indicator of the market reaction to the mineral right or the absence thereof. Sales of property with intact mineral rights can apportion a value to the mineral right by uses paired sales, income based adjustments, or polling. This would be the basis for making an adjustment for mineral rights or the absence thereof.

Income Methods

Income techniques can be applied to determine a contribution of a mineral right. A gross income multiplier can be used to estimate the mineral value based upon the initial bonus value times a typical Gross Income Multiplier. Traditionally that multiplier has ranged from 2.5 to 5, with 5 usually applied to properties very near production and the lower numbers being further from production. Discounted Cash Flow of rents is applicable.

Indirect Methods

Polling of industry experts, petroleum landmen, title lawyers, the National Association of Royalty Owners website (www.naro-us.org) and local persons who have leased their minerals or they have knowledge of other minerals leased by family members was used to estimate the current market lease rates for mineral rights.

Another indirect method is applicable. In states with pooling (a. k. a. - integration or unitization) Oil & Gas Commissions frequently holds hearings to integrate unleased parties into a drilling unit. The applications contain the applicant's efforts to lease the unleased parties and the offers made. The most recent hearing minutes can be read to determine the amount of bonus money paid to unleased parties being integrated by the Commission.

In states without integration (a. k. a. Force Pooling, Pooling, Unitization) or that don't go through the same procedures, it may take considerable effort to obtain much information. In states like Arkansas, Texas, Colorado, this information is typically available on line on the Oil & Gas Commission website.

"Appraising" vs "Apportioning" Minerals

Appraising a severed mineral interest involves finding other severed interests to compare with the subject. Extracting sales info from un-severed sales is far less reliable. This is because the mineral buyer has no interest in the surface. The surface buyer has an interest in the mineral as a concern over the control of the property.

Without the mineral right, the owner might not be able to prevent development nor would they be able to dictate well locations to the driller. They have lost executive control. This could have a dramatic impact upon properties such as dude ranches where the rural lifestyle might be disrupted by the noise of a rig or compressor. It could impact the poultry farm where noise might frighten the chickens or cause hens to stop laying eggs.

There are very few instances where mineral rights in oil prospective areas are valueless but in dormant areas, marginal areas, untested basins, etc., the value may be so low as to be statistically invisible. How do you segregate $10, or even $50 per acre from the fee simple as an allocation of mineral value? It is very difficult and usually unnecessary.

Examples of the Unit Rule

Unleased Oil & Gas Potential

The landowners agricultural value may be dominant and will be appraised with consideration to the added value from speculative production.

Leased Land for Hunting

The leasehold interest for hunting may be secondary to the value of the land for Grazing or agriculture.

Unitary Holding

Is there a unity of ownership and highest and best use? Is there contiguity between the properties? For instance, would a parcel set aside to protect the Ivory Billed Woodpecker be accessible to a drilling rig? Would the site be accessible by horizontal drilling or exception? Would the extra cost affect the value of the mineral holding?

Division vs Apportionment

Apportionment relates the relative value of the various estates (sticks in the bundle). These are dissimilar types such as leasehold vs royalty interest; surface rights, etc. You must weight the dominant stick in the bundle. This can prove to be no easy task.

From United States v. Tishman Realty and Constr. Co.

"The fact . . . that a valuation reached has in it baffling elements of speculation and surmise does not mean that it should not be employed. One guess may be better than another guess, since not all guesses have in them the same element of intelligence. The realization that a considerable amount ofiop; conjecture is involved should not paralyze the function of deciding, but it should induce humility. Dogmatism is clearly out of order in a modern

valuation case."

Dogmatic valuations of mineral rights are not to be expected. Rapidly changing prices for oil and gas makes those valuations a moving target and change occurs much more rapidly than the overlying surface rights. In conventional real estate rents and prices do not tend to change very fast except for some dramatic market move like the Great Recession of 2008. Such moves can be local such as when a one-industry town sees the major employer close.

In mineral rights, gas and oil prices are generally felt nationwide. While regional differences occur, prices fluctuate from coast to coast and can drop 30% in a few weeks. In 2008 oil prices had doubled in barely eighteen months only to crash in August ending the year lower by two-thirds. These price fluctuations seem to occur is much shorter cycles. Significant abrupt changes in prices has occurred at least ten times in the past twenty years. Some dips and spikes occurred only in natural gas while others seemed to have impacted only oil, or often, both oil and natural gas.

Risk Adjustments

Value relates to proximity to production. Geologically that statement is suspect since wells side by side can be radically different. But in general, the public cannot distinguish the difference. Therefore, the royalty buyer, who like the landowner, is not privy to the trade secrets of the oil companies.

Production Categories

Proved Producing	- Production that is on-going from an existing well
Proved Shut-n	- Production that is shut in but proved
Proved Behind Pipe	- Production in an existing well where well logs indicate will produce
Proved Undeveloped	- known as "PUD" - it is fairly certain

that it will produce but has not been drilled. The SEC allows companies to book those reserves for five years. They must develop them within the five years or take them off their books

Probable Reserves- Geologically similar to producing areas and believed potentially producing

Possible Reserves - Wildcat acreage far from production (say five miles??), aka "Goat Pasture" - a term widely used by landmen and geologists alike to denote areas where few or no wells are producing and where wildcat drilling to date has produced no favorable results.

Typically the appraiser will "risk" potential future production similar to below -

Producing	100% of remaining reserves
Behind Pipe	80% of remaining reserves
Proven - Undeveloped	50 - 60% of remaining reserves
Probable Reserves	10-15% of remaining reserves

Chapter V - Discounting Partial Interests

In solving a problem, the thing is to be able to reason backward- Sherlock Holmes

Quick Review of Fractional Interests

Fractional or Partial Interests "are divided or undivided rights in real estate that is less than the whole". Basically, this is an interest held as tenancy in common and owners are co-tenants. It applies to stock in property ownership agreement, simple title or a partial title such as water rights, mineral rights or an easement.

Dissatisfied Partners

Co-tenants who are dissatisfied, can sell their fractional interest, seek to partition the property, or ask for partition by a court. Such interests are less fungible than simple title controlled by one entity. Courts have ruled on discounts from zero to 90%

When a Discount is not Normally Applicable

Discount issues are usually raised when no contract specifying division of the property exists or where heirs disagree over the sale of property. The courts are usually reluctant to apply discounts in family situations, divorce, or where the land can easily be partitioned, especially if the land is raw land with no improvements. This is generally the issue with mineral rights. A divided mineral interest will sell as readily as a one that is undivided.

Discount for Partial Interests

Is there a discount for partial interests? If the mineral exploitation affects the value, then a discount to the surface value is appropriate. This is situational appraising. You have to use judgment. Whether mineral rights or not are involved, discounts are highly subjective.

But the real question is what is the impact of a severed mineral estate from a discounting point of view. Is there a discount to the mineral right, especially if split [i.e.- 50% of mineral rights reserved as an example]?

Impacts upon Value

Does the split affect the value? Mineral rights leased out are usually a passive interest. The mineral owner is not a working interest partner and has no executive rights anyway. In most situations, there is no reason to make an allowance for discounting the split estate of minerals.

Persons with less than 50% do not have lease executive power in some jurisdictions. ROYALTY deeds convey only a part of the mineral interest and may need to be discounted. Partial interest deductions are rarely necessary.

Discounts are Highly Variable

It is difficult to determine the appropriate discount, if any, that is applicable to a given property. Mineral rights are very often divided. Mineral rights, however, do not seem to be often affected by minority interest except in the case of Royalty Deeds.

Royalty Deed vs Mineral Deed

Royalty deeds do not convey executive rights and while the owner may receive all royalty income, the owner of the mineral right still sets agreements for royalty percentage and bonus money. Royalty deeds are to be avoided. Royalty deeds one place that they can be useful is a sort of substitute for a life estate. If a landowner wishes to transfer property to a child or heir such as a farm but retain the income from existing gas or oil production, then the royalty deed with a reversion is a means to do so.

Discounting Fee Simple for Environmental Concerns

You may have to adjust the value of surface if the property is impacted by the lack of mineral rights. This relates to the problems created by development of minerals rather than the loss of income from that development. Who wants a rock quarry in their back pasture? Or an old "one-lung" gas powered engine sans muffler running a pump jack.

Noise issues are being addressed in some states. La Plata Co., CO (Durango area) has passed noise ordinances requiring gas wells to keep noise levels down. The noise is from pumpjacks disposing of water and gas compressors required to put the gas into a high pressure transmission land. They also have screening and set back regulations. Painting tanks and pump jacks earth tone colors also reduces the visual impact of equipment.

To the industry it seems odd that people complain about pump jacks and well heads that are visible from the road, yet no one appears to complain about telephone equipment, electric power lines, nor cell towers in a similar fashion.

The latest controversy is over the issue of "home rule" where towns have attempted to ban all drilling, particularly their own ill-conceived definition of fracking. It appears the public attributes the entire drilling, production and transportation processes to something they call fracking. Fracking, a corruption of the term hydraulic fracturing, existed since the 1940s. Prior to that "fracking" consisted of lowering a torpedo tube full of nitroglycerine down the hole and igniting or "shooting" the well. In a few instances, a premature detonation fracked the shooter as well. A ban on fracking is a ban on drilling. And when landowners have minerals under a town (minerals often retained long before the town expanded over them) then a taking has occurred with such a ban.

Courts generally have taken a dim view of such bans despite public sympathy for same. A taking could trigger suits that would force a municipality to pay thousands of dollars per acre for mineral rights that underlie the city and create a financial crisis for the town. Denton, Texas voted just such a ban and by one estimate will have to pay landowners at least one billion dollars.[11]

Dealing with the Mineral Right

If you disclaim the mineral rights, the impacts on the fee

---

[11] Professor John Baen of North Texas University stated such in the local Denton news.

simple may be substantial and when you encounter the mineral rights on a property reserved, you should point out that the surface owner has less than 100% control of his property. The URAR checkbox does not have a check off for 'fee in surface'. For many this is a new paradigm. The appraisal of fee simple tracts has changed for the foreseeable future. You will not be able to ignore it long. You will have to learn how to value mineral rights as part of the assignment. And when disclaiming mineral, appropriate commentary needs to be made.

Chapter VI - Methods & Strategies

"In the past five years, the price of oil has ranged from ten cents to thirteen dollars and fifty cents per barrel." - Professor William Wright, 1865

Approaches

Cost Approach

The Cost Approach is never applicable for mineral rights as real property. The wellhead and ancillary items related to drilling and production are owned by the operator. The passive interest of the mineral owner therefore, is in the real property. Minerals are real property, but once produced oil and gas become personal property. In a working interest, or over-riding royalty, the interest is better described as an intangible business interest. Its value relates to the production of the personal property. The cash flow becomes the focus rather than the remaining life.

Sales Comparison Approach

Direct Sales of Mineral Rights

Private parties sell both producing and non-producing property on a per acre basis. Typically the seller is a private party and the buyer is a royalty company or oil company. Occasionally, royalty companies put together large tracts for resale to the oil company and as part of the deal they may keep a over-riding royalty interest in the properties. Since mineral interests can be transferred even when there is no production, this is the best way to value mineral rights where no production has been established there or nearby.

Decline Curves and Reserve Estimates

A simple truth about oil and gas is that the product declines with production. Since a oil reservoir can contain gas, oil, and water plus impurities (salts being the most common and usually dissolved in the formation water) as gas and oil are extracted, increased amounts of water will take up the space thus vacated.

This decline in production can be plotted over time and the result is a curved line. Since it is usually a "log-normal distribution", decline can be plotted on log normal paper where the X axis is linear and the Y axis is natural logarithm.

Formal math was introduced about the time of World War II. Arps[12] (1944) introduced a formula still used today. Based on empirical studies, Arps equations first suggested that there were three curves; the straight line exponential curve, the curved hyperbolic curve, and the harmonic curve. The harmonic is a special curve. There is little need to understand anything in depth except production declines as product is produced, and that it usually declines in a systematic and predictable fashion. That predictability can be translated into an estimate of the total expected ultimate recovery (EUR) and from the total already produced, an estimate of the remaining reserves can be determined.

Fetkovich[13] refined Arps formulations with the notion that the exponential decline is actually a reflection of how gas in a single state diffuses from a closed reservoir. In other words, gas escapes the reservoir in a predictable fashion just as air escapes a basketball at a predictable rate. Arps empirical curves were a reflection of more than mere curve fitting. There was a thermodynamic reason gas wells depleted in predictable fashion.

The appraiser needs not to be overly concerned about decline analysis. A competent estimator, like competent appraisers, will not produce identical results and for the same reasons. Each parameter which requires some judgment on the part of the appraiser or estimator is an area where divergence of opinions appear. Do not fear this, simply expect it.

---

[12] Arps, J. J.:"Analysis of Decline Curves", Trans. AIME (1944) 160, 228-47

[13] Fetkovich, M. J.: "Decline Curve Analysis Using Type Curves", Jour. Of Petrol. Technology (June 1980). This was typical of several of his publications in the 1970s and 1980s.

The precision of estimating future production is limited by

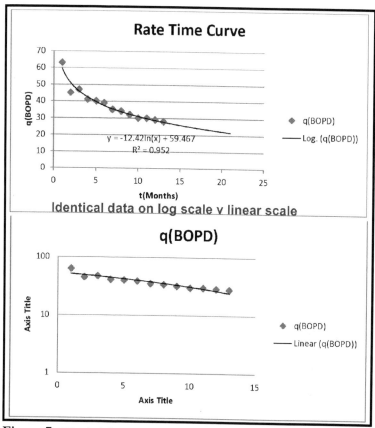

Figure 7 - Identical data based on linear scaling versus logarithmic scaling.

the geology. That probably won't be your biggest challenge. Even assuming that three different estimators might be 10% or even 30% different in estimated reserves, the future price of natural gas is even less predictable and the appraiser may be required to make a decision about price that would have a larger impact upon future prices than does the actual reserves.

The basic reason to use an estimator is to reduce your risk. This outside expert, if they make a mistake at least is not your mistake. If it is wrong, then you cannot be held accountable for it at some level.

Sales Approach from the sale of gas reserves in ground

This usually involves comparisons on a per unit basis. $ per 1,000 CF (MCF) of gas or $ per bbl of oil for producing property. After calculating a reserve estimate, the company will make an offer based upon the estimated volume. Alternatively, a company may offer a field or large tract of mineral leases held by production for sale to another oil company at a rate based on the remaining reserves. They may convert the gas into a equivalent amount of oil on a BTU or price basis and report as BOE (Barrels of Oil Equivalent.)

For instance, ABC company may sell XYZ company a stake in their oil field in the Eagle Ford. They sell 40 wells and 10,000 acres with the following projections. They estimated the reserves per well that remain to be 90 BCF of gas and 400,000 bbl of oil. The estimated that the undeveloped portion of the acreage holds 1,000,000 bbl of PUD (proven but undeveloped) reserves. The sale price is $160,000,000.

90 BCF is 90,000,000 salable MCF units. The PUD reserves are calculated at 50% for reserve risk. Thus, 400,000 + 50% of 1 million = 900,000 BOE. 900,000 BOE x $28 = $25,200,000 for the oil; and, 90,000,000 x $1.50 = $135,000,000 (160,200,000 total) or converting MCF to BOE using 12 as a multiple (which is less than the price ratio and more than the BTU ratio). Thus, the BOE of the gas is 7,500,000 and add that to the oil would be a total of 8.4 million BOE. That is about $19 per BOE.

To demonstrate the inexact nature of this process, let's reverse the numbers to represent MCFGVE. 900,000 BOE x 12 = 10,800,000 MCF (10.8 BCF) add to 90,000,000 = 100,800,000 x $1.50 = $151,200,000. Or, considering the sale price as fixed, $160,000,000 ÷ 100,800,000 MCF = $1.587.

Although different, the difference is small compared to the uncertainty that surround price. Price fluctuations during a full year often exceed 30% of an benchmark price. The average price over a year will be different from the average price on January 1st or June 30th. But the average of the year may be close to the average of those two dates.

Finally, this method actually is a blended mix of income and sales method. And in this author's opinion, it is frequently the most reasonable estimate. The biggest issue is to determine the value of the product in-the-ground. And there is a simple solution when you don't have sources. It is a general observation that the industry frequently applies the multiplier of 3 and in the case of oil or gas priced in the ground, one-third the market price is the price in the ground.

Income Methods

Gross Income or Net Income Multipliers

Gross incomes annually are multiplied to estimate a market value. Many taxing authorities rely upon this method. Oklahoma values estates at seven times the previous years income (which usually leads to over-valuation.) But typically that three multiplier comes in again. Why three?

Since oil and gas are wasting assets that necessarily take years to deplete, there is no reversionary value usually and the time value of money implies a very high capitalization rate. Investors expectations often cited a simple truism. I want the return OF my dollar, a return ON my dollar and a dollar for the risk of investing. $1 invested, $3 returned. Even today, many oil investors want to recoup their investment within 18 months and no less than two years. With unconventional wells, the bulk of the total reserves will be exhausted from a well within the first two years of production. It will be the income from the "out years" that are the actual profits.

A weak well many never return all of the capital that is invested in it. The very best wells can pay out in three months but may have to make up of the weaker sisters. Even though there is an expectation that shale plays are blanket deposits which implies a uniformity in the formations and production, there are a lot of geological and engineering issues that can impact the production.

In some areas, encountering an unexpected fault can result in not being about to complete a well, or formation water may simply overwhelm the borehole.

When a well makes excessive amounts of salt water, that water has to be re-injected on or near the site, or taken to a disposal well. Disposal (injection) wells re-inject fluids back into deep formations, often into formations deeper than the fluid originally came from. This presents two problems. If the salinity of the fluids is different then the fluid has to be either diluted or concentrated to make it similar. Otherwise, an electrochemical reaction can cause corrosion or plugging of the well pore spaces as salt attempts to precipitate from the mixture. The second problem relates to quakes. Please refer to Appendix B for an expanded discussion of the subject.

Discounted Cash Flow

A multiple of DCF techniques are applicable to valuing any income stream whether passive lease income or working interest or production. There are variations of the technique but with all discounted cash flows selection of an appropriate capitalization rate is a must. Unlike building and commercial properties where the property is presumed to have some value at the end of an investment period, oil and gas production has a high capitalization rate. The risk is high, and the value at the end of the DCF is low or may even be zero.

The length of the discounting should be based on the expected life of the well, but years beyond 10 show reduced net present values and at some point can safely be ignored for the most part. Thus, most DCF tables will only estimate from 10 to 20 years even if the field life is anticipated to be 30 years.

The Assignment

When a person receives an assignment the scope of work has to be agreed upon. The appraiser, however, may need to modify that scope once field work establishes what and how the value is impacted by nuances of the agreed problem to solve. In other words, identify the problem.

In most city assignments, developing oil and gas was never a big issue as the difficulty of siting without disturbing the houses was problematic. Companies avoided those areas. But with horizontal drilling, a rig can stand off a mile from a particular spot and drill and develop gas from beneath commercial districts and residential subdivisions without disturbing the peace and quiet above it.

But both rural and urban property may or may not be a true "fee simple" property. In reality, few properties are fee simple. The fee simple is encumbered with utility easements, use restrictions, zoning, deed restriction, and a host of other impediments to unfettered use of the property. Standard Rule 1-1 [c] (iv) requires the appraiser to identify "any known easements, restrictions, etc." It is incumbent upon the appraiser to identify and address these issues. Often the issue is simply addressed by observation that the issue is typical or is common in that market and has no market impact. Further, the appraiser must identify the real property interest being valued.

Identify the Problem

Avoid disclaimers when possible. Consider the impact on value of the mineral interest whether included or not. Correctly identify the rights appraised. Correctly determine highest and best use and the dominant estate. Understand the unitary holding and its impact upon value. Avoid hypotheticals or speculation.

Since lenders can be less than understanding about mineral rights, it is necessary to educate them or at least, clearly explain any issue you have encountered. It isn't necessarily a big deal, but handled wrong the Appraisal Management Company (AMC), whose employee may be in Mumbai, India, will get giving the appraiser fits as they try to pound the one inch square peg into the one inch round hole.

The problem is only compounded in rural property. Here the appraiser will need to know if the mineral rights are intact or not when the area is being developed for oil and gas. You will also have to make a judgment about whether or not the property is being negatively impacted by development at the very moment the local economy starts heating up and prices are rising.

In a town situation where the property is not leased and no development for oil and gas is likely in the foreseeable future, it will be difficult to apportion any value to the mineral rights. Absent or not, the value of the property is the same. If the deed actually reserves the mineral rights then the proper term is "fee in surface". That is actually what you are appraising, not the fee simple. Since appraisal forms don't have that option, an * is appropriate - some commentary upon the actual rights appraised. Again, anticipate questions from underwriters.

Do you include the mineral right or not? If you exclude a mineral right that is clearly there, then you are doing an appraisal of a partial estate - not fee simple.

Frequently the situation arises where the appraiser refuses to include the mineral right in the valuation of an estate that has a lease or production. This is probably not doing your client a service. By excluding the minerals, the appraisers foists the responsibility of the appraisal to a third party. The result is you appraise the fee simple (less minerals) and someone else appraises the mineral rights. The sum of those two values may be higher than the value of the whole estate rights. This may have tax implications for the heirs of an estate.

Remember that by appraising less than the fee simple, you can violate the Unit Rule which is unacceptable under USFLA regulated appraisals such as a federal taking or condemnation. The solution is to associate with a qualified resource appraiser so that you can value the entire fee simple estate including the mineral rights before proceeding with the appraisal.

Sample Statements

The Fee Simple estate is appraised. In the absence of minerals the estate appraised is fee in surface.

For areas with valuable minerals

Fee simple includes all the rights of ownership. The rights appraised herein are fee in surface and do not include any contribution from the mineral rights. The absence of mineral rights may impact the value of the overall property however, we have used similar sales that have no mineral rights contribution or have made an adjustment to the land value from paired sales.

The surface rights are valued and do not include the mineral right. Development of nearby properties for oil and gas [or coal, metals, etc.] may impact the value in the future but currently land values are driven by their value to buyers for [agricultural or residential] uses.

Other statements that might apply.

The rights appraised are those of the fee in surface and do not include the mineral interest. This constitutes a partial interest in the estate. However, that interest is the typical way in which properties transfer and producing mineral rights are generally severed from the fee simple title. This is typical and no adjustment is required to the comparables because they are equally impacted.

If the Minerals are Lease

The situation is complicated once a mineral right is leased. The minerals are now **leased fee** whereas the surface is not. And if the lease does not explicitly prevent a drill rig from being placed, a portion of the surface can be used to construction a well pad. Such drilling pads were temporary and small. After a well is drilled the area actually affected could be well under an acre.

Today horizontal rigs may have four or more wells on a single drilling pad. The pad and attendant compressors and tanks could take up two acres up to five acres or more. It becomes a substantial industrial site with potential for high truck traffic, noise, and lights at night. It is unsightly to eyes accustomed to the pastoral scene that it once was.

Even if you disclaim the mineral right, the impact upon value of a nearby pump jack must be taken into account even if you determine the impact to be zero. Correctly addressing the leased fee issue and impacts should show the appraiser is knowledgeable on the subject and should increase the clients confidence that the appraiser is credible. Unfortunately in the current banking climate, they often treat competency as a vice, not a virtue.

Including Mineral Rights in the Value

Non-Producing Wildcat Acreage

Non-producing and non-leased minerals likely may have very little value.

Wildcat but Active Leasing

This is where either mineral sales or a weak secondary method of multiplying the lease bonus currently offered in that area times three. Landmen use that method frequently and buyers will often make offers of 2.5 to 3 x the lease bonus. Be sure the lease bonus amount, like rent surveys, are current.

Developed /Drilled but no Check Yet

The appraiser should consider using an outsider to value the mineral contribution. In lieu of that, the appraiser should be acquainted with some advanced methods of dealing with the producing property.

Hiring the Expert

Remember that merely hiring an engineer or geologist does not solve your problem. The estimator must understand the role and limitations of using professional assistance. If the hired professional understands appraisals then you are at a distinct advantage. But even if they claim to, it does not hurt to go into some detail with the expert as to your goal and the scope of their work. The typical appraiser without geological or petroleum background may wish to have the estimator calculate (project) a reserves estimate with present worth values for you.

Communicate the Results

Don't try to complicate the issues. Explanatory text needs to be simple and well supported. A narrative is often preferable but form appraisals such as the 1004 can have addenda.

Networking with Other Appraisers

Transparency in some of these transactions will be difficult to determine but exchanging information with other appraisers will result in far more data and data arm's the appraiser to do battle with the dark forces of the marketplace.

Report mineral sales and intact fee sales where you can identify mineral value to me at www.roxnoil.com and I post a list of sales and information provided. Be sure to include location, consideration, acreage (net acres of minerals, too.), buyer, seller, date, and book and page. Any information about the conditions of sale are extremely important.

Sources of Information

The State Oil and Gas Commission
> Hearing information for pooling requests usually contains lease offers made during negotiations. In Texas the regulatory body is the Texas Railroad Commission, Oklahoma has the Corporation Commission. In Oklahoma, production figures for gas wells are held there, but oil production is handled by the state Tax Commission.

The Trades
> Petroleum Trade Publications
> The Petroleum Technology Council
> NARO chat room and other similar sites
> SPEE (Society of Petroleum Evaluation Engineers)
> The local Coffee Shop regulars
>> If you can separate the wheat from the chaff
>
> Your peers who work with mineral rights
> Landmen and Oil company personnel

Farm Services Mineral Appraisal forms

They are wholly inadequate for the appraisal of mineral rights but FSA appraisers have long used the forms and the instructions that used to accompany the form was misleading in the worst way. Instead of declining production, it suggested the income stream be averaged from past performance which almost always guarantees an inflated estimate.

Solution

Do not use their form as a guide to valuing minerals, rather value the minerals then fill out the form.

Speculative Value

One method that has been applied in court and sadly, has even won in court is a method of speculating upon some future value of the mineral which currently is untested and undeveloped. It is not market value. The assumptions necessary are so low probability that the entire procedure is simply a fanciful scenario with little basis in fact.

Assume you have a bare field miles from production. It is what some call goat pasture. What is its value? It's value is that of similar undeveloped goat pastures. It is a sales approach problem.

Now pretend that it has some mineral value. You estimate that the chances are 50% that a well will be drilled and successfully find oil - i.e.- they either drill or not, 50/50. You must project the ultimate production zone, the success of drilling and completion. And project a future production level, a future expense level, and a future royalty and bonus payment. Then stuff all that into a discount cash flow spread sheet and a number will come out. Is that value? Without testing, geological mapping, nor even simple seismic testing, is that a reasonable assumption?

Is it credible to create a series of hypothetical conditions and extraordinary assumptions about the subject in order to apply the method described above? The oil companies themselves may run these scenarios in order to rank prospects upon their potential using the information that they have. They do not call this market value. It is Expected Monetary Value (EMV) and serves little more than to identify which prospects are the most likely to be profitable.

For estimating market value, this is counting your chickens before they hatch and thinking the egg is worth more because it will become a chicken some day. It is a scenario that is lacks support and is considered speculative which renders it outside a traditional market value method.

Surprisingly, however, there have been instances in the courts where plaintiffs prevailed on the issue of making damage claims on the basis of just such arguments. It is still poor practice and speculative.

Part II

Valuing the Mineral Right

Chapter VII - An Engineering Valuation

Engineers like to solve problems. If there are no problems handily available, they will create their own... -Scott Adams

Engineering Value v. Market Value

    A mineral interest has a business enterprise value to the oil company. The owners of the leasehold are seeking income. They anticipate value from future production. The value therefore, becomes an economic issue for them. Can they expect to invest in the project and to sell more product than the cost of finding, drilling, and producing the stuff. This analysis is based on the Expected Market Value (EMV) which is a probabilistic projection. Sometimes it is called "Market Value" but generally, it lacks the formal analysis of market sales. It is an income projection.

    Engineers and geologists make these forecasts well in advance of drilling. They do so to estimate the potential for the economic extraction of oil or gas (or other minerals). Geologists early in the analysis tend to be optimistic so as to sell their deal to investors or management. A buyer may hire another geologist to vet this analysis and typically, it may show a more conservative approach. These are scenario analyses.

    Appendix I has a sample evaluation that might be prepared by an engineer or geologist. While it estimates a value, typically, it should not be mistaken for an appraisal. It would make excellent support, however, for estimating the contribution of the mineral right either as part of fee simple, or as a separate value of mineral rights. You probably will note that in the discounted cash flow engineers tend to use a fixed 10% discount rate (which is the mandated discounting found in SEC 10-K filings). And they will refer to the results as "Present Worth" rather than "Present Value". They mean the same thing but avoids the word "value". In all cases, it generally will refer to the discounted value of the product in the ground. That value will be discounted in two ways.

    First, the reserve estimator will project the EUR (expected ultimate recovery) and then an adjustment is applied to the reserve estimate. It is based upon the risk that the calculation is wrong, mechanical failure can occur, or when projecting by proxy what future PUD (proven but undeveloped) wells will do. They may not meet expectations.

The second discounting is the time value of money and the risk that prices will change. SPEE annual surveys give projections of discounting from polling results, and if presented with sales evidence, the discount rate can be calculated.

Market Value relates to the value of the rights to that product. It will reflect the estimate of reserves, and the time discount above. But the less tangible aspect of market value is the reaction of the buyers and sellers. When a play is new, there is limited evidence to support the engineered value. It is rather speculative. But buyers and sellers, anticipate good things in the future and will be ahead of the actual results that determine whether that value is high, low, or spot-on.

Once a new play has developed and matured, which is typically in three to five years, the market value reverts to the norm. High prices are not justified by the engineering results and those same results support the value from being discounted heavily. It is during those early years that production results need to be confirmed and the best way to do that is by vetting sales.

A Comparison of Fee Simple and Severed Estate

	Fee Simple	Severed Mineral Estate
The Estate		
The Players	Landowners, usually local	Petroleum professionals
The Market	The local land market	National mineral buyers
Marketing Time	Takes on the land M. T.	Usually <90 days
Value Sought	Market Value	Investment (?) Value

Types of Mineral Appraisals

Market Approach	Engineering Approach
Deterministic always	Probabilistic usually
Present Value	Present Worth
Weights Sales Approach	Ignores Sales Approach
Considers Income app.	Relies upon Income App.
Uses Simple GIM	Uses DCF frequently

Both Market and Engineering Appraisals must consider risk

Chapter VIII - Gathering the Data

*Errors using inadequate data are much less than those using no data at all -
Charles Babbage*

The most important part of valuing the mineral right is the unglamourous job of collecting all the information and sorting into a useable and reasonably accurate data set. It also is USPAP friendly to comment upon the quality, quantity and accuracy of the data collected.

It is a sad commentary upon the way people handle their finances but the truth is a lot of mineral owners carry a box full of check stubs to their CPA, usually in April, and let the CPA sort it out. The CPA does this - for a price, or simply uses the 1099 royalty income statements to figure the tax and leaves the rest up to the owner. No one audits these documents for accuracy.

And often, the owners have only a modest clue as to the nature of the interest they own. They were offered a lease and they signed on the dotted line. The lease may include the land they live on but as often as not it was a long-sold family estate where someone had the foresight to retain the mineral interest. That, in turn may be divided among heirs.

Now, a drilling company proposes a drilling unit. The land retained is included in the drilling unit. But the actual well proposed is a "cross - section" well. It will only be partially upon the subject lands. What is the actual ownership in each well as drilled?

Chapter IX - Methods for Valuing Mineral Rights

The meek shall inherit the Earth, but not its mineral rights- J. Paul Getty

Minerals without Production

Deeded Sales

In Arkansas and Oklahoma, mineral deeds must affix deed stamps to the deed. In Arkansas the rate is $3.30 per $1000 of transaction value[14]. In Oklahoma it is $1.50 per $1000. If the deed is explicit enough to tell you how many acres is in the tract, then you can calculate the per acre (net mineral acre) value.

Courthouses often keep mineral deeds in the Oil & Gas book separate from regular deeds and where leases are also recorded. Ask the clerk for assistance. Locating these deeds in the courthouse varies from county to county, not just state to state. Most are on line but many small counties have very little on line. You may be thumbing through books for the date range you are interested in page by page.

You want to find the deeds closest to your subject and the most recent sales. It is also good to have explicit descriptions of the interest. A vague description will mislead you. An example of an explicit description might be:

"It is the Grantor's intention to convey 12.80 net mineral acres of the above described parcel to the Grantee. Grantor and Grantee hereby acknowledge the conveyance is for 12.80 acres."

Armed with this information, the appraiser has a clear idea of the scope of the transaction. Virtually all of these transactions are cash and most are arm's length. The only question being the motivations and expertise of the seller and buyer. The private

---

[14]Such fees can change at the whim of the legislature. The rate has been changed in the past so if doing a retrospective appraisal, ask the county clerk what the documentary stamp rate was then.

seller, if an elderly person who needed to go to the rest home, or had no heirs, may have sold the property without proper professional advise or was forced into a quick sale. There is no substitute for talking to a participant in the transaction and therein lies the problem.

In states without deed stamps the only source of information is to contact the buyer or the seller. If the buyer or seller is a royalty company, they are unlikely to disclose exact numbers. They may provide a few details but rarely the price. In a moment of weakness, perhaps the per acre price. The seller, if a private party is more likely to disclose but many consider that personal information and can take umbrage upon being asked. And that is assuming you can even locate the seller. That can be a difficult task in itself. If you explain WHY you want that information as well as the who you are, you will be more likely to get results assuming you do locate those sellers. I generally start with something like,

> I am a real estate appraiser and am valuing an estate in Van Buren County and discovered a deed in the courthouse indicating you sold 40 acres of mineral rights to Blue Goose Royalty Co. Since I need comparable sales of mineral rights to value my client's property, I was hoping to ask nearby mineral sellers how much they received for their minerals, so I can get an estimate of the contribution of mineral rights to my subject. Would you be willing to tell me how much you were paid for your mineral rights?

No one likes cold calling strangers to ask about personal financial issues, but there isn't a good alternative if you don't have at a minimum, the documentary stamps and the actual amount of mineral rights sold. Even for states with disclosure rules, mineral deeds were not considered part of the deeds. In 2006, this was the situation in Arkansas until two appraisers approached a state legislator and asked him to ask the attorney general for an opinion on whether deed stamps applied to mineral deeds or not. His answer was yes and from that point on most counties have complied. It is also obvious that a number of deeds recorded, particularly in the south part of the state, do not contain documentary stamps and are almost certainly not complying with the law. When Chesapeake Energy sold its interests in the Fayetteville Shale, White County, Arkansas, alone, affixed over $200,000 of deed stamps to the transaction with BHP Billiton. Mineral sales do contribute considerable money into the state coffers.

Multipliers

The traditional landman method of value mineral rights is to multiply the current going rate for leases to a multiple, usually 3. In "hot" areas, that multiple might be 5. Currently some oil plays are as high as 7 and this relates to product price. Low prices means low multiples. Old wells with low production also bring low multiples. If a lease is in effect, then the multiplier may also depend upon how much the royalty is. A 1/8th royalty might be a multiple of 4 where a 3/16th might be 4 and a 20% royalty might command a multiple of 5.

Local lease prices can be discovered by asking people recently leased or when oil companies petition the Oil & Gas Commission[15] for a drilling permit under unitization (force-pooling), they must state what they will offer anyone being force-pooled. This varies from area to area but keeping as close to the subject as possible is a plus.

The appraiser can poll area residents or establish a relationship with a local landman. Once they understand what you are doing and why you want the information, they can become good sources of information. Try to reciprocate with information useful to them.

One last source of information are lease newsletters that compile information about rates in most active counties in the nation. Lierle Public Relations who publishes a national newsletter. Their website is http://usleasepricereport.com/

---

[15] This process varies from state to state. Most have either a set lease term, or the applicant testifies that they have offered the highest royalty and unit lease price in that drilling unit.

Example:
> Local sources tell you that current lease offers are from $500 - $600 per acre. In a current application before the Oil & Gas Commission, the applicant seeks to pool the unleased parties and states they will give $600 per acre which they swear is the highest unit value offered in the section (or unit). The Lierle U. S. Lease Price Report stated that this county had offers from $300 per acre up to $800 per acre. The appraiser believes the best estimate of lease is $600 and 3x that is $1,800.

It is your judgment as to the right figure to use, but in the absence of any other information, this may be the best method available for tracts that are not producing. And again, that x 3 factor may be the best figure to use for a multiplier.

An excellent source of lease and deed information is the local abstract company. If you can establish a relationship and make it plain you are valuing minerals, then often you can get general information. Many local abstractors also buy and sell property. They are often acting as brokers for mineral transactions.

Producing Property - No Reserves Estimate Methods

Early in the scope of work decision process, the appraiser needs to determine what methods are necessary and which can be omitted. If you are including the value of a old mineral interest in a property that is only a few dollars per month, is it economically feasible to run a full reserves estimate on it? Would the buyer do so?

Typically, in small parcels, with low potential for future development, the multiplier methods can be applied. This is how a buyer would approach it. If the income is only $100 a year and has been under $300 a year for the past 20 years, then the buyer would hardly pay more than a modest multiple of the annualized income.

Deeded sales

Deeded sales can be applied if sales are found in areas where production has been established. The issue is to compare apples to apples. If you are valuing a property where 20% of the reserves have been depleted, how do you adjust for a comparable where 40% of the reserves have been depleted? Is a sale of undeveloped minerals useful as a guide to your valuation except as a rough estimate of the top value?

Those kinds of questions plague attempts to value producing property by sales. In general, sales are the weakest method of valuing producing properties despite being the default method of valuing property without production. In most assignments, they are marginally useful.

Buyers, particularly small royalty buyers, are prone to seek unleased properties from owners who have so far refused to accept offers from the oil companies. When integrated, the process shows up in the public records or on records of compiled data - like Drilling Info, The Well Map, and IHS Petrodata.

While this data on line with the attendant software can be incredibly expensive, sometimes the local oil and gas commission records or the local geological log library or a university energy division may keep copies of IHS data which is published periodically. Drilling Info is more affordable.

Therefore, when valuing production that is more than a few months old, the deeds that are most commonly transferred are for land where drilling is about to commence or just commenced. The reservoir is untapped, but the property being valued may be depleted up to 50% or more. Making that adjustment will require you to estimate the remaining reserves of both the subject and the comparable sale.

Multipliers

If a property has an income history, then an offer may relate to the past production history. An offer might state they will give you the same amount of money that you have earn in the past 36 months or 48 months. This "months to payout" relates to the total amount already earned. If prices are rising, then this method favors the buyer but if prices are falling, it might favor the seller. Shale plays have presented a problem because the rate of decline is so high, buyers cannot give the sum of those high income years if the following years will see a serious decline in production. Therefore, they may offer only 24 months of cumulative income or maybe 28 months.

Years to Payout is a variation on multipliers and simply totals the past three years production income. This is a net income multiplier (before income taxes). Most check stubs will show deductions for post-production expenses including ad valorem taxes, severance taxes, and conservation taxes.

Producing Property - Reserves Based Methods

An engineer or trained professional can calculate the remaining reserves in the ground (under a unit or well that is in production.) This usually requires at least two or three years production history. These "decline curves" can then be vetted by a trained engineer or appraiser to estimate the amount of reserves remains. We refer to these experts as "estimators."

Some valuers do their own decline curves, however, by separating the estimating reserves process from the valuation, you create a buffer between the appraiser and liability for the reserve estimate. This is useful as a way of reducing your own professional liability.

The valuer then "risks" - applies a factor for risk - these reserve estimates and multiplies the owners interest (division of interest) to estimate the net reserves that the owner has. From that the owner can apply one of several methods.

In Ground Unit Pricing

Companies sell reserves in the ground. Again this is commonly considered to be around one-third the current market price. And time and again, good estimates of the reserves sold between companies frequently show transaction pricing that reflects that one-third market price. This can be estimated from sales by companies which are reported in trade papers or reported on the SEC's "Edgar" system for public companies. It is a disclosure requirement. Investment houses may track these prices based on the "M & A" - mergers and acquisitions of oil companies where they sometimes set a value for the reserves in the deal, or the investment manager estimates the price.

There are sources of this information. A magazine called "A & D Watch" published by Hart Publications, monitors the acquisitions and divestitures of oil and gas companies. These transactions frequently state the in round price estimated by either the participants or outsiders. It can be reported in either barrels of oil equivalent (BOE) or in MCF of oil equivalent (MCFE) but often is reporting the oil and gas reserves separately. Or it simply says the sale was valued at some number, such as $21 per barrel of oil reserves and $1.66 per MCF of gas reserves Likewise, Raymond James annually publishes information regarding transactions. In 2013 the average calculated price in ground from their research was $1.50 per MCF (gas) and $17.54 per bbl. (Oil). That is slightly less than a 12 ratio based on price between the price of gas and oil. We have previously noted that on a BTU basis the ratio should be 8.

The differential between the ratios may be explained by the cost of transportation. Oil from the Bakken (North Dakota), for instance, costs about $20 per bbl. just to transport by rail to the refiners in Texas. Natural gas is transported by pipeline and that is much cheaper. Also, oil is more commonly discounted for impurities and light oils, which with fracking, presently have glutted the market and are suffering discounts of 20-40% in many markets. Sellers prefer to lump these "liquid-rich" reserves with oil since it indicates a higher market value. At the time market prices for oil were between $90-105, then the one-third rule would seem to be not working. Once we discount the $20 for the transportation costs, the net price is dropped to $70 or so. That means one-third of 70 is $23.33 per barrel. And $1.50/MCF is near the $4.50 price of gas for NYMEX gas futures (2014 prices).

Another example in following the trades (oil and gas news publications) was the sale of Arkoma Basin assets by Forest Oil. The old Arkla Gas Company sold assets to Houston Gas and years later Houston sold out to Forest Oil. When Forest merged with Sabine Oil & Gas the decision was to get rid of these older wells. Those wells were sold and reported that they produced "22 Mmcfd of natural gas equivalent during the third quarter and had estimated proved reserves of 159 bcfe (billion cubit feet equivalent), 100% natural gas as of December 2013." The sale price was $185 million. Thus, $185,000,000 divided by 159 BCF = $1.16 per MCF (1,000 cu. ft. - which is the unit that natural gas is sold in.) At that time natural gas prices were hovering around $3.50 per MCF. $3.50 divided by $1.16 = 3.02, thus you see the reason for the 3 x multiplier.[16] This unit value method is the simplest way of valuing mineral reserves where reserves are calculated.

Discounted Cash Flow

In surveys of industry professionals over 80% of the respondents typically chose discounted cash flow (DCF) as the primary method of valuing property. Left unsaid is that this is predominately for large transactions where millions of dollars are at stake and teams of engineers and geologists have input on the parameters and basic estimates necessary to determine a value by the DCF method. And few respondents admitted that they relied totally upon DCF as the only way of valuing property. Most applied two or more other methods. For the small mineral owner the DCF method is likely used far less than with corporate transactions involving hundreds of acres or wells.

In small transactions, it is pretty clear that there is less reliance upon DCF and more reliable upon multipliers and $/BOE/MCFE. Some financial methods are also applied such as profit to investment ratios or internal return on investment (IRR).

As the most sophisticated methodology, discounted cash

---

[16] This may be on of the more difficult things for appraisers to overcome. A multiplier of 3 seems very low to appraisers more accustomed to multipliers of 9 and higher for commercial real estate and apartments. However, minerals are a wasting asset and therefore may have zero value at the end of the production cycle, or may not be viable again for decades. Once again the issue of risk means investors who are savvy to the ways of the oil patch know that market conditions can change almost overnight. Witness the 2014 meltdown of prices during the month of November. This is a repeat of the 2008 run up in the price of oil which then proceeded to collapse from a high of $140 to $50 or less and then rebounded to $70-80 per barrel. Only in the most recent and tempting new plays will the multiple of 4 or 5 be accurate.

flow has to consider the basic method to apply. Do you escalate or not? By that it is meant that during the length of years chosen to discount are you going to use the current price of oil and gas, or escalate annually? Will you average the annual production, or predict annualized production.

Arguments for the escalation method is that prices do change year to year and production changes year to year. Both are true but neither price nor production is precisely predictable. If the EUR is fixed (estimated by the decline curves) then allocating the decline is based on the estimated decline rate which is derived in that decline curve. If the curve suggests the well will decline to zero before the cut off year (say 10 years) then the DCF needs to be shortened to that figure. If production is longer than ten years, the logical cut off is the total life of the well, or the average of all wells in the field. Technically, it is most accurate to plot each well decline individually and sum the annualized estimates of the production. That can prove time-consuming not to mention mind-numbing. Or the computer can do it if you have the right program.

The NYMEX exchange has natural gas and oil price projections for years into the future. You can apply those figures but the figures should be adjusted. The local price of oil and gas products are normally not NYMEX futures prices. They are discounted by the local market factors which can be 30% or more. The local price can be vetted by observing the check stubs of the owners. The check stub will have that price on it. Comparing to NYMEX in the same time frame will give you an adjustment factor. Be aware that gas prices even within a single check stub can vary according to the quality and end sale terms of the gas or oil.

The alternative method is to use today's price and divide the EUR into even annual amounts consistent with the length of time being discounted. The results may differ somewhat but are likely not to be so different as to be unrecognizable. Anything past the tenth year is going to be discounted significantly. Likewise, is that projection any less likely to be correct than a projection where you honestly cannot predict what the price and production will actually be? Valuation rarely comes after complete development of a drilling unit (pooled unit.) There will be part of the wells that exist and part that have not yet been drilled. The undrilled sites must be considered. Discounting them as risk-adjusted reserves is one major step in that direction.

Hoskold's Premise, Other Methods

Several methods of estimating cash flow and discounting that cash flow for the time-value of money and for risk can be applied. All methods are estimates only and actual income can only be determined by actual production and the variability of the market prices which vary with supply and demand and economic conditions.

Hoskold's Premise is similar to the Inwood Premise and is rarely used for oil and gas properties. Hoskold's Premise is a sinking fund formula that is most often used in mines and quarry valuation. The unrisked rate (usually a treasury bond interest rate for a similar length as the term of the analysis) is added to a risk rate and allows for both yield and a return of the investment.

The formula accounts for the length of the investment and the T-bill interest rate is added to a risked rate. That risk rate varies but is often calculated at three times the T bill rate or perhaps as high as five times the safe rate.

This example also calculates the years to payout as present worth divided by the annual income. The formula provides for a return on the property in addition to a return of the investment.

Hoskold, a Swedish mining engineer, suggested that this technique might be appropriate for valuing mines where the value is reduced to zero as minerals are extracted but annual production can be maintained; thus funds have to be set aside to invest in a new mine once the minerals are totally depleted (the reversion equals zero). The application to oil and gas properties is complicated by the more volatile pricing mechanism vs. mines and the declining production of a well.

The use in intangible royalties relates to the way that investors view those oil and gas royalties. Most assume that at the end of the production, the reversionary value of the royalty is for all practical purposes zero. The spreadsheet above calculates the value of an investment where a safe rate and a risked rate are included in the sinking fund and the annualized income is anticipated to be relatively steady over the holding period until the investment terminates. That is, it is a wasting asset with no reversionary value.

Within the industry there are numerous metrics used as a proxy for market value. That includes the internal rate of return, profit to investment ratios, years to payout, etc. Within individual companies, they vet property using internal data unavailable to the public or at least not easily understood by the public.

Formation parameters can be estimated from electric logs. From those, the amount of pore space is estimated, and a determination of the oil and gas thickness, in addition to formation water can be determined mathematically. From that the volume of oil and gas is estimated and the percentage believed to be recoverable is estimated.

Other methods include estimating the in situ percentage of carbon and relating that to the total anticipated recovery of oil. This would actually involve coring a piece of the rock in the borehole and testing for its carbon content. If you see the term TOC (total organic carbon) then they may be referring to this method. Many companies rely upon this method as it can be kept internal as a trade secret and is not disclosed such as production records and electric logs are disclosed.

The valuer of oil and gas rarely has the opportunity to get this information nor the expertise to do anything with it if they did have access. It is clear that the way the engineer and the appraiser will handle valuation is different.

Summary

There are two styles of "market value determination. The engineer and oil company will analyze the data they have which is proprietary and comes at considerable expense. From this they will make a multi-disciplined judgment about the total potential reserves in a given prospect.

The company will generally seek valuation that is based upon the payout of the projected reserves using conservative estimates of success and price. This is a way to rank the various prospects against each other, and arrive at an Expected Monetary Value (EMV), which isn't necessarily a "market value".

Few companies are willing to proceed with a project which they don't think will return their investment within three years. Since they cannot control price, they project their numbers accordingly and usually will proceed with the most favorable prospects.

In this, the oil company will have utterly no concern over what the prices are being paid as sales of minerals. They evaluate on the bases of reserves and only reserves.

The landowner or mineral buyer has to deal with the uncertainty of production that is less a concern of the oil company and their engineered data. And they do concern themselves with what others are paying for the mineral rights. The offer made under the circumstance of having little more than prior production records or a proxy thereof, means the buyer may discount the offers based upon the amount of product already produced, or if the market prices are falling. They likewise may increase their offers in the face of stiff competition from other buyers, rising product prices, and increased drilling.

Often these buyers are anticipating value from the actions of the drillers and invariably pay more for actively drilled areas over that of the wildcat acreage.

Chapter X - Anatomy of the Play

Our ignorance is not so vast as our failure to use what we know - M. King Hubbert

The Boom - Bust Cycle

Petroleum is a boom and bust cyclical business. It has been since the days of Col. Edwin Drake. Drake would developed such innovations as well casing in addition to his early efforts at drilling, ended up broke. The price of oil went from $1 per gallon to 50¢ per barrel[17] (42 gallons - also known as a tierce).

In the early 1920s the fear of a shortage of oil lead to government efforts to encourage exploration. By the depression, the price of oil had crashed to 5¢ per barrel and states introduced proration, a process where producers were required to reduce the production level to a percentage of the wells capacity. This reduced the supply of oil over time. It discouraged drilling. The price of oil rebounded.

World War II brought a large need for oil and failure of a ready supply of fuel contributed to the downfall of both Germany and Japan. Development of the Middle East oil was given encouragement with President Dwight Eisenhower musing that using their oil saved our oil for the future. A drilling boom ensued during the 1950s and that resulted in driving oil prices down to a low in the early 1960s. America responded with gas guzzling automobiles and Arab countries watching their oil sell for less than $2.00 a bbl. were alarmed and this led to the creation of the Organization of Petroleum Exporting Countries (OPEC) in 1960. And the peak drilling activity of the late 1950s begin a long slide that was not stopped until 1972.

---

[17] Wooden caskets were 42 gallons from the time of King Richard III, who deemed a casket of 84 gallons as a wine puncheon and a half size barrel was called the tierce. It became the standard for dried fruits, oil, and dried fish. It weighed about 300 lb.

By the late 1960s the supply of natural gas was diminishing and a shortage was predicted. At that time, a long dismissed analysis by an obscure Shell Oil research geoscientist was given a second look. In 1955 M. King Hubbert had predicted that the U. S. was on a path to reach peak production in the early 1970s. He was proven correct when considering the conventionally developed reserves known at that time.

The OPEC oil embargo in 1972 saw panic in the U. S. Drillers ramped up and oil and gas exploration increased to a peak nine years later when the price of oil collapsed. It collapsed in part because of the amount of U. S. oil found and the removal of price controls. Oil prices had been depressed under a system of pricing "old oil" from "new oil". The end result was these new oil controls were actually creating a floor for OPEC production which kept constant pressure on the price of domestic crude. "Old Oil" wells were shut in and thousands of marginal (a.k.a. - stripper) wells were plugged and abandoned.

Dealing with the Rapidly Changing Market

In the early stages of a new play, landmen ("lease hounds") will be leasing heavily and paying more and more as the geologist and the drill rig prove up production. During these times, speculation is so great, valuing mineral rights is more guess than any rigorous valuation. After roughly two years, enough production data is available to make a better educated valuation. This is based upon the production and the remaining reserves. The appraiser must be cautious about valuing those early properties. Such speculative values have proven time and again to be far too optimistic. The uninformed may find that an oil company has purchased some mineral rights for say $10,000 per acre. But keep in mind that the oil company is obtaining 100% of the rights and the production. The landowner, on the other hand, has something less, perhaps as little as $12\frac{1}{2}\%$ of the total package. Exchanges between landowners and mineral buyers will rarely be so much as 50% of what an oil company may pay. Again, this likely reflects the gross income from that interest is greater to the oil company than to the landowner's leased fee.

Once this euphoria subsides then the sale of mineral rights

will level off. Buyers of mineral rights typically will attempt to purchase unleased properties that are near or in a drilling unit that is proposed to be developed. Oil companies usually have to petition (file application) to unitize (pool) an area. With that application, the explorer will list the unleased parties within the unit. In states without pooling by the state, this is not an issue as they will not have unleased parties. But they may attempt to lease others to include in the pool, or make a joint operating agreement with the lessee of those minerals, if already leased.

Thus, most sales of mineral rights will be in the early stages of development. But once development starts, purchases will typically relate to remaining reserves and based on the past income history of the property. There is one situation to avoid. Mineral Deed & Royalty sales of production from an existing well may reflect the low income of the property without considering the potential for future wells to be drilled in the unit. Many units can support 8 or more wells but only one well is required to hold the property by production (HBP). Therefore, uninformed sellers may sell mineral rights far below the market value.

So the early stealth lease hounds may lease at low bonus and low royalty but once the news is out that there is a play going on, the lease prices rise radically. From that point on, drilling will ramp up rapidly and as most booms go, it will be far above what is supportable over a long period of time. At some point, the rig count falls, prices stabilize, and as wells deplete, the value of the mineral right will be dictated by demand, remaining reserves, and oil and gas price pressure.

Appendix A - Engineering-Style Report

A Sample Demonstration Mineral Evaluation for the

Jed Clampett Estate
Section 16, 21, & 22 - T8N-R11W
Faulkner County, Arkansas

November 1, 2013

Prepared for:

Larry, Curly & Moe PLLC
101 Capitol Ave.
Little Rock, Arkansas

Prepared by

I. B. Valuer
Registered Professional Geologist

Mineral Evaluation for the
Jed Clampett Estate
Section 16, 21, & 22 - T8N-R11W
Faulkner County, Arkansas

At the request of Larry, Curly & Moe, attorneys for the Jed Clampett Estate, we did a study to estimate the value of minerals owned in sections Section 16, 21, & 22 - T8N-R11W, Faulkner County, Arkansas. The estate owns a total of 220 acres of minerals with 100 acres in section 16, 80 acres in Section 21 and 40 acres in Section 22 (see Exhibit I). These mineral are currently under an oil and gas lease to Mid-Continental Oil & Gas Company of Shreveport, LA through May 6, 2014. This lease is subject to a one-eighth (1/8th) royalty paid to the Clampett Estate on proceeds derived from production of any oil and gas that is produced and sold in any of the sections that the estate owns an interest. A deed was not made available and the Harry Larry, attorney provided a copy of the lease. I have assumed in this study that the Estate owns 100% of the mineral interest described in the lease.

The Fayetteville "play" is being developed as a natural gas resource producing from the Fayetteville Shale formation and produces only natural gas as an economic mineral. These minerals referenced above are anticipated to be developed in the near future. The Arkansas Oil & Gas Commission (AOGC) has granted two permits to drill horizontal test holes. Chesapeake Energy (now BHP Billiton aka BHP) has staked a well in Section 21 and Southwestern Energy Inc. (SEECO) has staked a test in section 22.

Future Production

The evaluation is based on production records obtained from th AOGC on line records. No production currently exists. To estimate future production, the area of study was extended to include currently producing natural gas wells. A total of 20 producing horizontal wells are located in the 8 N - 11 W township. A production decline curve was created from the production records of the existing wells in this township. From this curve we can estimate the total estimated ultimate reserves (EUR) of each well.

The EUR found in these wells ranged from 0.2 BCF (billion cubic feet) to approximately 3.5 BCF of gas with an average estimated EUR of 1.25 BCF per well. I chose the SEECO Jobe 8-11 #1-10H well located in Section 10 - T8N-R11W (see

Exhibit II) as the model well and applied this well as a proxy for the typical well in the subject sections. This decline rate is only an estimate and could vary significantly from future production in the subject area.

The mineral acreage located within the same section that has a producing well is considered to have proved undeveloped remaining reserves. Additional reserves added from drilling is considered to be proved undeveloped reserves if they have at least a 90% probability that the reserves will equal or exceed the reserve estimate.

Probable reserves are unproved reserves which are anticipated to be proved by development drilling in and reserves have at least a 50% probability of reaching or exceeding the estimate. Possible reserves are unproved reserves which have the geological characteristics to suggest productivity but are distant from production or have known geological features that differentiate the formation characteristics from that of the proven reserves, such as a fault. These reserves should have at least a 10% probability of reaching or exceeding the reserve estimate. The above classifications come from the SPE "Guidelines for Application of Petroleum Reserves Definitions" (October 1998).

The subject area has no production but offset production is found to the north and west in sections 8, 9, 10, and 20. Therefore, we classify the subject as unproven probable undeveloped reserves in the Fayetteville Shale. They have at least a probability of 50% to find reserves that will achieve an estimated EUR of 1.25 BCF or more.

Gas Prices

The gas price used for future production is based on the current NYMEX natural gas price and applies no annual increase to the price. The current price is $3.80 per thousand cubic feet (MCF). An additional deduction is made for local price and post-production expenses and we applied 20% for that and arrived at a revenue per MCF of $3.04 per MCF.

Royalty Interest

The net royalty interest (NRI) in each drilling unit (based on one unit per square mile section) was calculated by dividing the number of acres owned by the estate within each section by the total number of acres in the respective section and multiplying that by the royalty interest (1/8th). Thus, the mineral interest owned is as follows on next page.

Location	Acres	Unit Size	Royalty Rate	NRI[18]
Section 16	100	640	0.125	0.01953125
Section 21	80	640	0.125	0.015625
Section 22	40	640	0.125	0.0078125

Future drilling in each of these sections cannot be predicted and is in the control of the operators who have integrated each section. Gas prices, drilling expense and early well performance will control the speed and direction of development. We have projected that the maximum drilling development per integrated unit (section) is eight wells. This is based on a typical horizontal lateral length of 4,630' ± spaced 650' apart.

Future development is based on the probability of success and based on a value per net mineral acre. An economic run sheet was prepared for the one net royalty acre in each section (unit). We used the model well with the EUR of 1.25 BCF. A discount rate was applied at 15% and we arrived at a Net Present Worth based on the Royalty model presented as Exhibit III.

The value is based upon the percentage of the drilling unit that the owner controls times the royalty percentage. This is referred to as the Division of Interest and is the decimal ownership of the income stream that the owner should be paid after deductions and ad valorem and severance taxes.

Income taxes are not included. From the decline analysis for the proxy well (the Jobe), we estimated the annual production expected over the following 10 years. We projected the price on a current market price basis. No annual inflation rate is estimated. We have assumed that inflation and the increase in gas prices will be offsetting.

The development projection, EUR, gas prices, and drilling and completion costs are all estimates and are highly variable,

---

[18] Net Royalty Interest

therefore the value of the estates minerals listed below are only estimates and subject to future events no one can predict accurately.

Conclusions

The Jed Clampett Estate owns 220 acres in three sections. This mineral is undeveloped but in an area of proven production. A 50% discount is appropriate. An economic analysis of a proxy well in the same township suggests that the proven production value would be 1.25 BCF of natural gas and that the analysis in Exhibit III suggests the value per acre would be $3,440 before discounting for the risk of drilling and producing. The probability is that a given well would be either more or less productive than the proxy well, thus, a 50% discount is appropriate.

$3,440/acre
x 50% (as proven undeveloped)
$1,720 per acre

Net Present Worth

It is understood between both parties that all findings, calculations, and conclusions are strictly based on the geologist's interpretation and Terrel Shields shall not be held liable for any decision, business or personal; any liability for any tax or regulatory burden; nor the use by either the client or any third party based upon the opinion rendered in this study. This study is not a substitute for an appraisal and does not render an opinion of "market value."

Signed Nov. 1, 2014

Sincerely,

Valuer's Signature

Exhibit I - Area of Interest

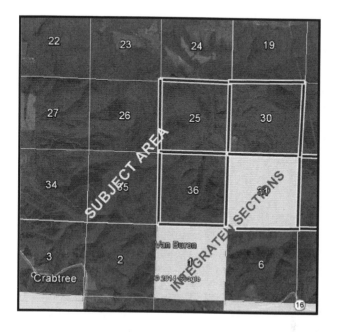

Exhibit II - Decline Curve of Subject Well

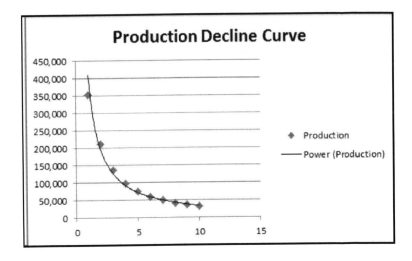

Exhibit III - Discounted Cash Flow

As proven undeveloped we have projected that a further discount of 50% is necessary

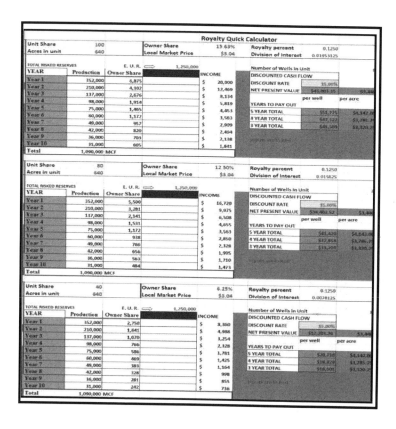

demonstration report only - 97

Appendix B - Fracking & Earthquakes

This is a big issue because it is something that impacts the public. Fracking (the actual hydraulic fracturing process) does not create these widespread quakes.[19] Applying the term "frack" to any and all oil field processes is an abuse of the word promoted by those whose view is that all fossil fuels should immediately be banned from use. There is a clear relationship between injection wells and small tremors.

The relationship of the disposal well (or injection well, if you prefer) and an quakes cannot be disputed. Not all disposal wells create quakes. In fact, large disposal wells in South Arkansas and Texas are quake free. regulate injection wells meant for oil and gas wastes.

All the studies I know of and the stated opinion of some government geologists is that earthquakes are mostly being caused by the injection wells. And, they usually occur in situations where the fluids are injected in deep formations where hidden faults are found. There are a lot of faults and its hard to describe but as each formation is lain down, there can be contemporary faults covered up and these may not be detected when geologists attempt to piece together the underground patchwork of geology.

Injection wells are vertical wells, not horizontal wells like those that are fracked. They often have very high volume pumps pumping into the formations as opposed to an actual frack which is a comparative low volume which is carefully injected in stages along the horizontal portion of the borehole. Should during that process, the fluids encounter an unknown fault, which is a line of weakness usually (very old faults may actually be cemented up by natural processes) then the fluids will not frack the shales but will travel down the line of weakness (and it does tend to travel downward into what is called "basement" rock.) In those cases, it usually ruins the well and the well borehole is cemented up and

---

[19]Fracking does create microseismic events that cannot be felt on the surface. Sensitive monitoring devices are lowered down existing wells and are continuously read during the hydraulic fracturing process. Rock fractures open up and the injected fluid containing a proppant (sand or glass beads) which when the pressure is removed, will hold open those fractures allowing the gas to escape to the borehole.

abandoned.

It is the injection wells that are the problem and even then, only a few places have a problem. Ohio (NE corner) seems to be one. There were a few very small quakes in the Barnett around Ft. Worth. Numerous quakes occurred near Guy, Arkansas. The initial injection well there was originally a wildcat well drilled in the 1970s. I actually consulted to one of the companies that leased this well and again, this well injected into formations that were much deeper than the Fayetteville shale. Further, that formation was very porous and thick. After studying the geology more, there is a consensus that the formation is thick because of a fault. Perhaps a reverse fault has slid the porous formation up over itself doubling the thickness and in the middle of that thick zone is the fault plane.

Interestingly the quakes in Guy, when plotted against time, started from the SW to the NE and at some point turned to the NW and made an upside down L pattern. This is almost certainly the shape of this fault block of porous sand and carbonates. In other words, the geological situation is conducive to earthquakes and even without the injection the occasional quake has been around. In the early 1950s a "swarm" of these quakes occurred along the Nemaha Ridge in Oklahoma that was unrelated to drilling (the state's biggest quake occurred then.) And near Guy, Arkansas in the mid-80s a swarm to the southeast only 10 miles shook for several years at a time when there was no drilling or production at all. It was called the Enola swarm. In other words, the geological situation is conducive to earthquakes.

I have no doubt the quakes in central Oklahoma to S. Kansas are caused by injection wells along the granite basement "Nemaha Ridge" - an ancient weakness that starts north of Ft. Worth, follows I-35 and then turns south of Wichita towards Kansas City. Along that ridge are lateral fault zones that were "relief" zones and what some Oklahoma City geologists believe is that the quakes are suspected to be along these lateral faults. I am not sure the solution but potentially they may need to inject into shallower zones at lower rates and stick to recharging old zones that were drilled and produced decades ago. The formation pressures there would be lower and should reduce the risk. The cost may rise as there is the issue of the potential for a rogue injection well operator to inject far above the pressure and volumes that their license allows for. That temptation is always going to be there if the financial incentive is great.

But the fact is that there are a lot of places fluids have been injected for decades without any quakes. South Arkansas, South Texas, West Texas, etc. have injected oil field fluids for decades without a single earthquake. Quakes are controlled by the geology and areas where injection is near faults that penetrate basement rocks in the right geological setting will potentially be a problem. Better siting of the injection wells is a potential solution.

However, there is no evidence that the process of FRACKING itself causes significant quakes. Once the initial controlled (stage by stage) frack is performed then the well is extracting fluids, not injecting them and so far, extraction does not seem to have much impact seismic activity.

Damages to a dwelling can be extensive or simply annoying. Quakes tend to flex a dwelling and in doing so, cracks rigid material. The wise appraiser will caveat the issue by proper disclaimers in the report.

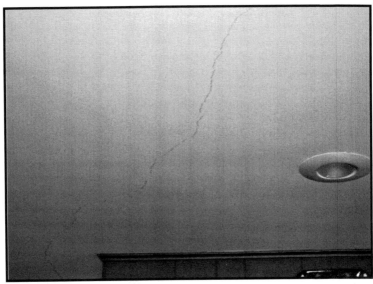

Figure 11- Cracked ceiling from a November 2014 quake in south-central Kansas. Photo- courtesy of Mark B. Winans of the Mark Winans Appraisal Group (316) 755-3448

Made in the USA
Middletown, DE
22 December 2015